Sustainable and Affordable Mobility for All: Putting the Heart Back into Technology

Sustainable and Affordable Mobility for All: Putting the Heart Back into Technology

Dr. Christopher Borroni-Bird

400 Commonwealth Drive
Warrendale, PA 15096-0001 USA
E-mail: CustomerService@sae.org
Phone: 877-606-7323 (inside USA and Canada)
              724-776-4970 (outside USA)
Fax: 724-776-0790

Copyright © 2025 SAE International. All rights reserved.

No part of this publication may be reproduced, stored in a retrieval system, transmitted, in any form or by any means, electronic, mechanical, photocopying, recording, or otherwise, or used for text and data mining, AI training, or similar technologies, without the prior written permission of SAE. For permission and licensing requests, contact SAE Permissions, 400 Commonwealth Drive, Warrendale, PA 15096-0001 USA; e-mail: copyright@sae.org; phone: 724-772-4028

Library of Congress Catalog Number 2025933342
http://dx.doi.org/10.4271/9781468609356

Information contained in this work has been obtained by SAE International from sources believed to be reliable. However, neither SAE International nor its authors guarantee the accuracy or completeness of any information published herein and neither SAE International nor its authors shall be responsible for any errors, omissions, or damages arising out of use of this information. This work is published with the understanding that SAE International and its authors are supplying information but are not attempting to render engineering or other professional services. If such services are required, the assistance of an appropriate professional should be sought.

ISBN-Print   978-1-4686-0934-9
ISBN-PDF   978-1-4686-0935-6
ISBN-epub   978-1-4686-0936-3

To purchase bulk quantities, please contact: SAE Customer Service

E-mail: CustomerService@sae.org
Phone: 877-606-7323 (inside USA and Canada)
              724-776-4970 (outside USA)
Fax: 724-776-0790

Visit the SAE International Bookstore at books.sae.org

Publisher
Sherry Dickinson Nigam

Product Manager
Amanda Zeidan

Production and Manufacturing Associate
Michelle Silberman

The GM Autonomy concept, with its electric "skateboard" platform foreshadowed the revolution in the automobile a decade in advance. Its next revolution, as discussed in this book, will be to reshape auto-mobility, enabling localized production of highly optimized vehicles for moving people and goods, that work in tandem with public transport.

Standing beside me,
From college to the present,
This book is for Laura

Contents

Foreword I. xi
Foreword II xv
Acknowledgments xvii
List of Acronyms xxi

Chapter 01 - Introduction
1.1. A Note on Definitions Used in the Book. 5

Chapter 02 - The Status of Global Auto-Mobility in 2025
2.1. Vehicle Usage 7
   2.1.1. Maslow's Needs Hierarchy—Applied to Cars 9
   2.1.2. Automobiles are "Overengineered". 11
2.2. The Rise of Automotive Electronics 16
   2.2.1. The Life of a Typical Car—a Link between Developed and Developing Worlds. 19
   2.2.2. Other Motorized Passenger Vehicles 23
2.3. Future Societal and Demographic Trends. 25
   2.3.1. Urbanization 26
   2.3.2. Age Demographics 29
   2.3.3. Obesity 31
References 38

Chapter 03 - The Recent Automotive (R)evolution 1990–2025
3.1. EVs. 44
3.2. Microvehicles. 55
3.3. EV Technology 57
3.4. Connectivity and Software-Defined Vehicles (SDVs). 62
3.5. AVs. 66

3.6. Autonomous, Connected, Electric Mobility.....................73
3.7. MaaS..75
References..86

Chapter 04 - The Emerging Automotive Landscape 2025–2035

4.1. Battery Performance...89
4.2. Battery Charging..92
4.3. Hydrogen Fuel Cells..99
4.4. Rethinking Chassis and Propulsion............................101
4.5. A New Automotive E/E Architecture..........................106
4.6. AVs..108
4.7. Cabin Experience..119
4.8. V2X..124
4.9. eVTOLs (Air Taxis)...128
References..132

Chapter 05 - Some Consequences of the Future Automotive Trajectory

5.1. Unintended Consequences.......................................135
5.2. Systems Thinking...139
5.3. EVs..140
 5.3.1. Environmental, Social, and Governance (ESG) Issues............141
 5.3.2. National Security Issues..142
 5.3.3. EV Mass Issues..145
5.4. Connected and SDVs..147
 5.4.1. Privacy..148
 5.4.2. Security...151
 5.4.3. Life Cycle Concerns..152
5.5. ADAS...154
 5.5.1. Cost...155
 5.5.2. Effectiveness...157

5.6. AVs..160
    5.6.1. Safety..160
    5.6.2. Accessibility..162
    5.6.3. Congestion..164
    5.6.4. Energy Usage..166
    5.6.5. Workforce and the Economy...............................169
5.7. Vehicle Affordability.......................................170
References..173

Chapter 06 - A New Paradigm for Auto-Mobility

6.1. The Automotive Technology Race..............................180
6.2. The "Inevitable" EV Transition..............................181
6.3. How Have Countries with Major Automakers Responded to the Rise of EVs?...186
6.4. Emerging Countries in the EV Value Chain....................189
6.5. How Did Traditional Automakers "Drop the Ball"?.............195
6.6. Traditional Automakers Need a New, "Sustainable" Strategy...198
6.7. Hierarchy of the New Framework..............................200
References..207

Chapter 07 - Rethinking Auto-Mobility

7.1. Urban Design for Mobility—the Ugly, the Bad, and the Good...209
7.2. Autonomy and Public Transport...............................220
7.3. MaaS..227
7.4. Removing "Friction" in a Vehicle Subscription Service.......231
7.5. Goods Transport...236
7.6. The Potential Role for City Government to Shape Future Mobility...242
References..244

Chapter 08 - Rethinking Vehicle Design and Development

8.1. Reducing Vehicle Energy Demand..............................248
8.2. Nature's Lessons for Future Vehicle Design..................252

8.3. Solar Power 256
    8.3.1. Right-Sizing the Vehicle 262
    8.3.2. Right-Sizing the Battery 266
8.4. Car-Free and/or Low-Speed Zones 270
8.5. The ACE Platform "Reference Design" Applied to Future Automobiles 273
8.6. The E-Kit as an ACE Platform for Nonmotorized Vehicles (NMVs) ... 279
References 290

Chapter 09 – Rethinking Vehicle Materials and Manufacturing

9.1. A Paradigm Shift 293
9.2. The Circular Economy 297
9.3. Natural Materials 301
9.4. Localized Production 305
9.5. Economic Benefits and Supportive Government Policies 314
9.6. An Opportunity for Automakers 323
References 326

Chapter 10 – Conclusion

10.1. Rethinking the Future of the Auto Industry 329
10.2. A Vision for Integrated Mobility 330
10.3. An Opportunity for the Auto Industry 331
10.4. A Sustainable Future for All 332
Afterword I 335
Afterword II 339
Index 341
About the Author 349

Foreword I

I am glad Dr. Christopher Borroni-Bird wrote this book! We have known each other since 2010, when we met for the first time at a future urban mobility colloquium in San Francisco, well before the hype around self-driving cars and all-encompassing mobility platforms. *Reinventing the Automobile: Personal Urban Mobility for the 21st Century*, the book he had co-authored not long before that, I had consumed with great interest. Naturally, I was excited and honored to be on the podium together, discussing what it would take to dramatically improve sustainability, safety, and equity in urban transportation. What followed for us was a regular thought exchange when we would discuss technical, societal, and at times philosophical matters around any kind of vehicle—those with four, three, two, and even without wheels [the latter being flying cars, aka vertical takeoff and landing (VTOL) vehicles].

Over the years, I have appreciated the opportunity to accompany a great thought leader on his professional journey from General Motors to Qualcomm and further to Waymo. Ultimately, he would follow his true calling and apply his knowledge, passion, and network to serving those in what is often called the developing or even underdeveloped world. I have often struggled with those categories myself, asking what it really means to be developed. In my own publications and lectures, I have typically cited the quote from Gustavo Petro, former mayor of Bogotá and current president of Colombia: "A developed country is not a place where the poor have cars. It's where the rich use public transportation."

While I still see some merit in this, I really appreciate the direction that *Sustainable and Affordable Mobility for All: Putting the Heart Back into Technology* takes when highlighting what we can learn from one another in terms of preferences and solutions for our transportation needs, be it in the Western world, in Latin America, sub-Saharan Africa, Southeast Asia, or elsewhere. In that sense, Chris offers with this book a great perspective as he discusses how the developed and developing worlds can learn from each other to turn around the unsustainable path many of us are on when it comes to our daily commutes, consumer habits, and travel patterns.

"Putting the Heart back into Technology" is one of his key messages, which underlines his passion as a scientist for mobility solutions based on even that—technology—and not mere business interests. Chris Borroni-Bird stays true to the socially responsible thinker and maker who offers his reflections on where the automotive industry is headed and what needs to be done to provide mobility for all; but also to ensure said sustainability, equity, and safety that we discussed passionately when we first met in 2010, Dr. Borroni-Bird pays much attention to what is happening in the real world as he provides quantitative data regarding our transportation habits and the resulting problems. Consequently, he is now making the case for right sizing, which is to design mobility solutions such that they meet the needs that we all have, no matter where we live: to get from A to B safely and sustainably, and to enjoy the ride, too. His call to action is to avoid overengineering in terms of size, mass, consumption, etc. To the contrary, it is to envision transportation solutions that are spacious and powerful enough to get us to our destination of choice, while remaining reasonable and sustainable to offer the privilege of personal mobility to everyone without ending up in an apocalypse.

When reading the manuscript, I was constantly reminded of a concept that I teach at the Stanford Business School as part of a strategy seminar: The Innovator's Dilemma, formulated by Clayton Christensen in the mid-1990s. The concept discusses how established firms are so focused on improving their existing products with every new model cycle, until critically neglecting customers who are satisfied with something that is just good enough. However, newcomers in the industry—often startups or cross-boundary disruptors—may seize an opportunity with seemingly inferior products, which ultimately not only catch on with consumers but may even create a new market that pushes out the incumbents: the disruption from below. In that sense, *Sustainable and Affordable Mobility for All* should be, at the very least, a reminder to decision-makers in the automotive industry that "good-enough" products might be growing up to compete with today's feature-laden vehicles, and it should serve even further as a recommendation of what can be done to avoid being disrupted. Managers, strategists, and ultimately everyone in the industry need to understand that we are not moving on a sustainable trajectory and that very soon there may be an inevitable imperative: "less is more." Chris Borroni-Bird not only provides convincing facts, he also knits all those together in a broader context, creating a fascinating story.

Once again, I am glad he wrote this book; it is the right book at the right time. Highly recommended.

Dr. Sven Beiker
Managing Director at Silicon Valley Mobility
—Palo Alto, CA, January 2025

Foreword II

I need a bigger toolbox. I am a transportation planner with 30 years of experience with virtually all transportation modes around the world. I have planned and designed streets, public transit systems, stations, and airports. I have planned and designed transportation networks for major technology campuses in Silicon Valley, traffic calming measures in Brooklyn, and major stations in Hong Kong. My job is to ensure convenient access while also building vibrant and attractive urban places. My tools are limited. The private car dominates our transportation network. But this comes with congestion, car crashes, and unfriendly streets that are hostile to humans on foot, on bikes, and using wheelchairs.

The passenger car has evolved into a blunt instrument. Standards, consumer preferences, and commercial realities have guided manufacturers to sell cars that are nearly uniformly large and heavy (Seats five! Plenty of cargo space! Powerful engine!), have long ranges (Large gas tank! Bigger batteries!), and are built from environment-damaging, non-renewable materials (Large touchscreen! New car smell!). But cars sit unused most of the time or are most often driven with fewer people, much less cargo, and for shorter distances than their specifications. This contributes to the negative externalities of cars: environmental degradation, traffic congestion, the construction of divisive highways, massive amounts of land dedicated to parking, and tragic crashes—often with vulnerable road users such as pedestrians and cyclists.

Despite the negative impacts, transportation, including passenger cars, remains the lifeblood of our society. Access to a car enables upward economic mobility. Cars provide critical access to opportunities including jobs, school, health care, and activities such as shopping, social visits, and recreation.

Some things have radically improved over my career, including vehicle safety features and improved fuel efficiency. However, stubborn traditional challenges remain. I have long been frustrated by the relatively small number of tools in my toolbox and the lack of transformative innovation. Transportation is hard. We need big thinking.

Chris Borroni-Bird is a rare big thinker. I met Chris at an event focused on autonomous vehicles (AVs) and radical change over 10 years ago. The 2015 book he co-authored, *Reinventing the Automobile*, was a revelation to me. Its ideas challenged the traditional paradigms of passenger cars and their impacts on our communities and our world. This visionary work inspired me to think bigger about the challenges, and the solutions, in my work. Borroni-Bird ups the ante in this book.
He combines his unique perspective of deep technical expertise and extensive experience in the automotive industry with holistic thinking, global perspective, and creativity.

Sustainable and Affordable Mobility for All presents a revolutionary but realistic vision for a new automotive industry and transportation system. It is expansive, covering everything from manufacturing to urban integration to regulation and operations. This vision addresses our biggest challenges including economic development, environmental sustainability, and universal accessibility. It presents ideas that are not only scalable, but also flexible to be adapted to local needs and contexts. It is informed by 10 years of rapid innovation and real-world experience: ubiquitous ride-sharing platforms, electric vehicles (EVs) in the mainstream, and robotaxis in revenue service to the public in major cities.

The ideas presented in this book can inspire a new vision for urban mobility to transform the private car from a blunt instrument to a precision set of solutions. It is a thoughtful mashup of technology, transportation demand and supply, human factors, and environmental sustainability systems, presented through a systems-thinking lens. As a transportation professional, I look forward to adding these tools to my toolbox!

Will Baumgardner, PE TE
Transportation Planner and Engineer

Acknowledgments

I am deeply grateful to SAE for embracing the creation of this book and to Sherry Nigam and Amanda Zeidan for their invaluable support throughout the entire process of writing this book. I wish to thank Sven Beiker, a respected automotive expert in Silicon Valley, for encouraging me to write this book and for writing a Foreword. For their endorsement and varying perspectives on public transport and globalization that enrich this work, I also want to extend my sincere gratitude to Will Baumgardner, Inderveer Singh, and Ricardo Apaez. I was also fortunate to have received insightful feedback on draft manuscripts from Roger Lanctot, Chris Groessbeck, Partha Goswami, and Sam Abuelsamid. Sam also graciously provided several images for the book, as did Steve Fershtl, Lincoln Wamae, and GM.

My career journey has been shaped by many mentors and colleagues. I am indebted to Sir David King, who later went on to be Chief Scientific Advisor to Prime Minister Tony Blair during the outbreak of "mad cow" (BSE) disease and advocated strongly for action on climate change. He (Sir David King, not Tony Blair!) supervised me during my PhD at King's College, Cambridge and provided me with a daunting research challenge with no guarantee of success, but one that taught me the importance of perseverance and luck! The photo shows me in 1989 beside the first machine I designed and developed. It was the first to demonstrate how much heat is released when gas molecules adsorb onto metallic surfaces, which is important for helping to design better heterogeneous catalysts such as those used in vehicle exhaust systems and in chemical plants for producing ammonia and synthetic fuels.

My life was then immeasurably changed by the late Tom Moore who took a risk and brought me across the sea to Chrysler in Detroit to fathom out what the National Labs were doing that might help the auto industry back in 1992, shortly after the Cold War ended and the Labs needed a new mission supporting US industry. He also encouraged me to evolve from a research scientist into a vehicle program manager, leading the gasoline fuel cell vehicle efforts in order to overcome hydrogen storage and infrastructure issues.

This led me to think about how technologies could enable future vehicle design, and not just be used for meeting environmental and safety standards. In 2000, Dr. Larry Burns at GM gave me the opportunity to develop innovative new EV concepts like Autonomy and EN-V that fused technology and design in the service of societal goals. These vehicles were the first to demonstrate the "skateboard" EV architecture and the fusion of autonomy, connectivity, and electrification, with EN-V having a form factor designed to support the theme of the 2010 Shanghai World Expo. His sponsorship of these programs inside GM was essential to them seeing the light of day. I was very fortunate to be supported by Mohsen Shabana and Pri Mudalige, who are both very practical engineers with a strong research ethos and, just as importantly, were always very positive and a joy to work with.

After anticipating that the automobile was becoming a "smartphone on wheels," I moved from the auto side to the tech side and found myself at Qualcomm and Waymo (part of Alphabet) working on the cutting edge of automotive technologies, such as vehicle-to-pedestrian communications, wireless EV charging, and robotaxi development.

In parallel with my career, I have been fortunate to receive support on understanding mobility needs at the other end of the wealth spectrum, in rural Africa. I would like to thank Professor Mark Bryden at Iowa State University for introducing me and my daughters to village life in Mali in 2009 and 2010. It was on these trips that I became curious to create mobility solutions for the developing world that would, in some way, leverage my work developing future vehicles. The thinking has evolved from creating complete solar-powered EVs to the point where the e-kit is a two-wheel "skateboard" that can retrofit to existing non-motorized vehicles (those that are pushed, pulled, or pedaled). I would also like to thank the Toyota Mobility Foundation for their vision in sponsoring mobility fact-finding visits to Zimbabwe and Kenya in 2020, where I was joined by a former GM colleague and friend, Richard Saad.

Finally, my heart goes out to Laura and our three wonderful children (Sophia, Christina and, Marco), for supporting me throughout all this. Thank you.

List of Acronyms

2W - Two-Wheeled (vehicle)

2W - Two-Wheeler (bicycle or motorcycle)

3W - Three-Wheeled (vehicle)

3W - Three-Wheeler (tuk-tuk or rickshaw)

AAA - American Automobile Association

AAM - Advanced Air Mobility

ABS - Anti-Lock Braking System

AC - Alternating Current

ACC - Adaptive Cruise Control

ACE - Autonomous, Connected, Electric

ADA - Americans with Disabilities Act

ADAS - Advanced Driver Assistance System

AEB - Automatic Emergency Braking

AGV - Automated Guided Vehicles

AI - Artificial Intelligence

AV - Autonomous Vehicle

AWD - All-Wheel Drive

BEV - Battery Electric Vehicle

BMS - Battery Management System

BOM - Bill of Materials

BRT - Bus Rapid Transit

BSM - Blindspot Monitoring

BYD - Build Your Dreams

CAAC - Civil Aviation Administration of China

CAFÉ - Corporate Average Fuel Economy

CARB - California Air Resources Board

CKD - Complete Knock-Down (for assembly)

CNN - Convolutional Neural Network

CPU - Central Processing Unit

CTP - Cell-to-Pack

DARPA - Defense Advanced Research Projects Agency

DC - Direct Current

DMS - Driver Monitoring System

DRC - Democratic Republic of the Congo

EASA - European Union Aviation Safety Agency

ECU - Electronic Control Unit

E/E - Electrical and Electronics

EN-V - Electric Networked Vehicle

EPA - Environmental Protection Agency

EREV - Extended Range Electric Vehicle

ESC - Electronic Stability Control

ESG - Environmental, Social, and Governance

EV - Electric Vehicle

eVTOL - Electric Vertical Take Off and Landing (aircraft)

E2W - Electric Two-Wheeler

E3W - Electric Three-Wheeler

FAA - Federal Aviation Authority

FCEV - Fuel Cell Electric Vehicle

FCW - Forward Collision Warning

FMVSS - Federal Motor Vehicle Safety Standards

FSD - Full Self Driving (Tesla)

FTC - Federal Trade Commission

GREET - Greenhouse Gases, Regulated Emissions and Energy Use in Technologies

HDPE - High-Density Polyethylene

HEV - Hybrid Electric Vehicle

HMI - Human Machine Interface

HVAC - Heating, Ventilation and Air Conditioning (or the Climate Control)

HWD - Holographic Windshield Display

ICCT - International Council on Clean Transportation

ICE - Internal Combustion Engine

ICEV - Internal Combustion Engine Vehicle

IEA - International Energy Agency

IIHS - Insurance Institute for Highway Safety

III - Insurance Information Institute

IoT - Internet of Things

IRS - Internal Revenue Service

JV - Joint Venture

L0-5 - Levels 0-5 (autonomy)

LCO - Lithium Carbonate (a lithium-ion battery chemistry)

LDW - Lane Departure Warning

LEZ - Low Emission Zone

LFP - Lithium Iron Phosphate (a lithium-ion battery chemistry)

LLM - Large Language Model (AI)

LTO - Lithium Titanate (a lithium-ion battery chemistry)

MaaS - Mobility as a Service

MTA - Metropolitan Transport Authority (in New York)

NCA - Nickel Cobalt Aluminum (a lithium-ion battery chemistry)

NEV - Neighborhood Electric Vehicle

NEXTCR - Next Generation Energy Technologies for Connected and Automated On-Road Vehicles

NHTSA - National Highway Traffic Safety Administration (a US government agency)

NiHM - Nickel Metal Hydride

NMC - Nickel Manganese Cobalt (a lithium-ion battery chemistry)

NMV - Nonmotorized Vehicle

NVH - Noise, Vibration and Harshness

OBD - On-Board Diagnostics

OEM - Original Equipment Manufacturer (typically the automaker)

OPEC - Organization of the Petroleum Exporting Countries

OTA - Over-the-Air (software updates)

PEMFC - Proton Exchange Membrane Fuel Cell

PET - Polyethylene Terephthalate

PHEV - Plug-In Hybrid Electric Vehicle

PHV - Plug-In Hybrid Vehicle

PM - Permanent Magnet

PMT - Passenger Miles Traveled

PRT - Personal Rapid Transit

SDV - Software-Defined Vehicle

SLI - Starting, Lighting and Ignition (battery)

SUV - Sport Utility Vehicle

TAM - Total Addressable Market

TCS - Traction Control System

TfL - Transport for London

TOD - Transit-Oriented Development

TNC - Transportation Network Company (e.g., Uber, Lyft)

UAE - United Arab Emirates

UE - User Experience

UK - United Kingdom

UN - United Nations

US - United States

USDOT - United States Department of Transportation

VMT - Vehicle Miles Traveled

V2C - Vehicle-to-Cyclist (wireless communications)

V2G - Vehicle-to-Grid (wireless communications for EVs)

V2H - Vehicle-to-Home (wireless communications)

V2I - Vehicle-to-Infrastructure (wireless communications)

V2L - Vehicle-to-Load (wireless communications for EVs)

V2P - Vehicle-to-Pedestrian, or Person (wireless communications)

V2N - Vehicle-to-Network

V2V - Vehicle-to-Vehicle (wireless communications and EV charging)

V2X - Vehicle-to-Anything (wireless communications)

WAV - Wheelchair Accessible Vehicle

WEVC - Wireless Electric Vehicle Charging

WHO - World Health Organization

WLTP - World Harmonized Light Vehicles Test Procedure

XUAR - Xinjiang Uyghur Autonomous Region

xEV - All Types of EVs (BEV, HEV, PHEV, EREV, FCEV)

ZEV - Zero Emissions Vehicle

ZTL - Limited Traffic Zone (Paris)

Chapter 01

Introduction

Mobility provides access to people, places, and goods. On a personal level, it improves quality of life, while at the national level, it stimulates economic development. Throughout most of human history, right-sized vehicles, such as the horse and bicycle, were the norm, to be supplemented by public transport solutions like the stagecoach, train, tram, and bus. Around 1900, cars and planes began appearing and, over time, addressed different distance needs. By the mid-twentieth century, highways drove a "need" or expectation for longer-range, higher-speed operation, which also drove increased demands for ride comfort and crashworthiness. Market segmentation created cars, vans, and trucks, with sporty and luxury versions, but it can be argued that all these vehicles are overengineered for typical usage and the vast majority of trips. These vehicles may last for 20 years and can end their life in the developing world, which also has many motorbikes and nonmotorized vehicles for serving the poorest people (those with incomes below $2/day).

Whereas early automotive pioneers, such as Henry Ford, focused on making vehicles affordable, current automakers seem more interested in developing features, such as hands-free driving (while still having to pay attention to the road), or trying to monetize "heated seats" with subscriptions or software updates. The developed world seems to be going down the wrong path because vehicles are becoming more expensive to purchase, repair, and insure. They are also becoming taller and heavier, and more dangerous to other road users. Emerging, highly integrated vehicle structures for accommodating the battery could make them more likely to be "totaled" and more difficult to recycle. The auto industry is on a path towards serving the wealthiest 20% of the population in the wealthiest 20% of countries.

Remote working, more durable and expensive vehicles, and more ride-hailing in an increasingly urban population could lead to fewer new vehicle sales in the developed world. Conversely, rapid population and economic growth in the developing world could lead to more new vehicles being sold globally each year. The path we are on could be disastrous for the planet if the developing world copies the developed world (as most wealthy people in the developing world have until now). However, there is room for cautious optimism because, out of necessity, the developing world has leapfrogged the West in terms of telecom infrastructure and mobile payments and could be the first to embrace innovative technologies like solar panel vehicle roofs.

I have been lucky enough to hold technology leadership positions at several companies, working on the cutting edge of vehicle autonomy (Waymo), connectivity (Qualcomm), and electrification (GM), while also having the opportunity to experience firsthand the challenges of moving people and goods in off-grid sub-Saharan Africa. This combination of automotive technology experience and global mindset began while leading GM's 2030 vision of personal urban mobility for *the 2010 Shanghai World Expo* and, simultaneously, volunteering in a rural Mali village on solar lighting, cookstoves, and water pumps. This book's hopeful thesis is that the developed world can learn from the developing world how to use energy and materials sparingly and provide more affordable solutions, while the developing world can benefit from the advanced technologies being created in the developed world that promise to be cleaner and more efficient than existing solutions. Everybody in the world should have access to affordable mobility even if it might drive new energy and materials usage, but it can and should be done in a way that is sustainable for the planet and affordable for everyone, including the poorest people.

The book is structured in a way that describes the current global mobility situation in Chapter 2 before covering electric vehicle (EV), connected vehicle, and autonomous vehicle (AV) innovation over the last 30 years (Chapter 3), and the emerging future solutions that are being developed that will reshape mobility, driven largely by automotive and technology companies (Chapter 4). Importantly, these same enablers can also be applied to micromobility and public transport. It should come as no surprise that there will be several issues created from these new solutions, because "any solution, at scale, creates a new set of problems," just as when the car displaced the horse-drawn carriage and seemed to make the city center cleaner by eliminating horse manure from the city streets. These issues and unintended consequences are discussed in Chapter 5 and are likely to be significant. For example, energy usage is likely to increase, not because future technologies are inefficient, but because they could induce more demand for mobility. Not only will this offset some of the potential environmental benefits, but the solutions that are being developed may not even be affordable for most people around the world.

To democratize mobility in a way that makes it affordable to everyone on the planet, this book pivots to making the case for a new approach in Chapter 6, which starts with the customer's needs (not the technology or commercial interests) and focuses on auto-mobility. I am using the term "auto-mobility" to emphasize the merging of the automotive and mobility landscape, where the latter includes micromobility and public transport. The same technologies of electrification, autonomy, and connectivity that are transforming the automobile can also be used across the broader mobility sector, facilitating greater integration between the two domains, and replacing the expensive automobile ownership model with affordable access to all kinds of mobility, as covered in Chapter 7. This will put the heart back into technology and give most people what they want—affordable and convenient access to people, places, and things.

For the vehicles themselves, which will still be needed, Chapter 8 shows how these same technologies can be leveraged to enable right-sizing, which will further reduce energy and materials needs, while making vehicles more affordable. Once the vehicle design is right-sized, it will be easier to apply circular economy principles and increase the use of natural materials and solar energy, as well as to stimulate local, community-level vehicle manufacturing production in many places, the focus of Chapter 9 (**Figure 1.1**).

Figure 1.1 Thematic flow for this book.

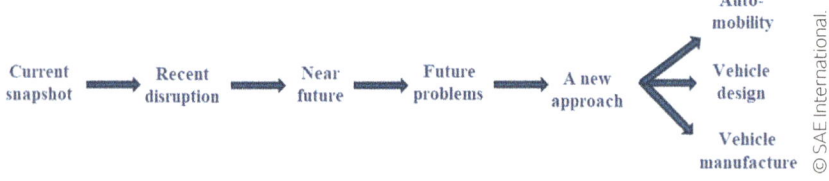

Of course, these proposed solutions will sit alongside the existing business model for many years to come, rather than replacing it, but I believe that this framework can be applied to the developed world, emerging economies, and even to rural, off-grid subsistence farming populations in places like sub-Saharan Africa and India that have many synergies with "last-mile" mobility needs across the whole world. Cities or regions are likely to adopt various aspects of the framework and to localize the solution to their specific needs. By doing this, we will be putting the heart back into technology by placing society and the people at the center and enabling everybody in the world to have access to mobility and to have more control over their local mobility system, instead of relying on multinational automakers, tech companies, electricity utilities, oil companies, and central governments. However, these major actors do have a promising opportunity to play a constructive role in developing these solutions. Traditional automakers, particularly, may need to rethink their current business model and the existential competition they face from newcomers, particularly from China's EV makers. I hope that they will want to develop new auto-mobility solutions that can leverage their expertise to create new products, markets, and even public–private partnerships while enabling sustainable and affordable mobility for all.

1.1.
A Note on Definitions Used in the Book

The terms "vehicle," "automobile," and "car" are often used interchangeably in this book as they are in everyday life. My aim is for the context to make it clear whether I am referring to automobiles or to two-wheeled and three-wheeled vehicles. In addition, vehicles that you and I drive may sometimes be called "personal," "private," or "passenger" vehicles.

I will use the term EV when discussing battery EVs (BEVs) even though the technical community prefers the acronym BEV to distinguish it from HEV (Hybrid EV), PHEV (Plug-in Hybrid EV), and FCEV (Fuel Cell EV). This book is intended for a broader audience, and most people refer to BEVs as EVs.

I will also use the phrase "autonomous vehicle" instead of the more accurate technical term "fully automated vehicle," for the same reason that it is commonly used by the public and in the media. AVs are not currently able to operate without human assistance (because they either need a safety driver or a remote operator), and so they are not truly autonomous but they are highly automated.

Claudio Soldi/Shutterstock.com.

Chapter | 02

The Status of Global Auto-Mobility in 2025

This chapter of the book is intended to provide a snapshot of how vehicles are used today and why they should be considered "over-engineered," despite their increasing use of lightweight materials and electronics. It then discusses how societal trends are expected to influence the way future vehicles will be designed.

2.1. Vehicle Usage

Let us consider how much the typical automobile is "overengineered" by considering data on how much of its capability is typically being used. This includes not only the vehicle's physical size and weight but also how much time it is used, and the average speed and distance driven each day.

In 1977, the US average vehicle occupancy was 1.9 persons per vehicle [2.1]. By 2019, this had dropped to 1.5 and for commuting it is even lower, at 1.13 [2.2]. Given that an average new car weighs around 4300 lb, this means that the typical passenger-to-vehicle-weight ratio is in the 5–7% range [2.3]. It may come as a surprise, but this is a similar ratio to a half-full 35,000 lb highway bus. By comparison, a typical 20 lb bicycle has a passenger-to-vehicle-weight ratio of around 700% [2.4].

According to a study by the American Automobile Association (AAA) Foundation for Traffic Safety, the average American driver spends approximately 17,600 minutes (or 293 hours) per year behind the wheel [2.5]. This translates to approximately 48 minutes of driving per day on average, meaning that it is parked for more than 23 hours each day.

The average traffic speed in most major city centers is around 10 mph, and has barely changed in more than a century. For example, according to a study by the New York City Department of Transportation, the average speed of vehicles in Midtown Manhattan is below 5 mph during peak hours and around 10 mph for the whole of Manhattan [2.6]. Numerous studies in London, Tokyo, Beijing, and other major global cities have shown similar traffic speeds [2.7, 2.8], and increasing this traffic speed is one of the motivations for cities wanting to introduce a congestion charge. Going outside the dense city center, the average speed of vehicles increases, and according to TomTom, across the whole day in Los Angeles, it is around 20 mph [2.9].

The average annual mileage driven by car owners in the United States (US) is around 11,000 miles according to data from the Federal Highway Administration, or just over 30 miles per day on average [2.10]. People living in rural areas tend to drive about five miles more per day on average compared to people living in urban areas because they often have less access to public transport and need to travel further to access schools, work, places of worship, hospitals, and so on [2.10]. The national average annual mileage is significantly higher than the average in many other countries, due in part to the vast size of the country and the prevalence of car culture (**Figure 2.1**). In contrast, 94% of trips in the United Kingdom (UK) are shorter than 25 miles [2.11], and the average annual mileage in the UK tends to be just over 7000 miles [2.12]. In China, the average annual mileage driven seems to be similar to the US, at 11,000 miles [2.13].

Figure 2.1 Passenger miles traveled (PMT) each day, 2022 National Household Travel Survey [2.10].

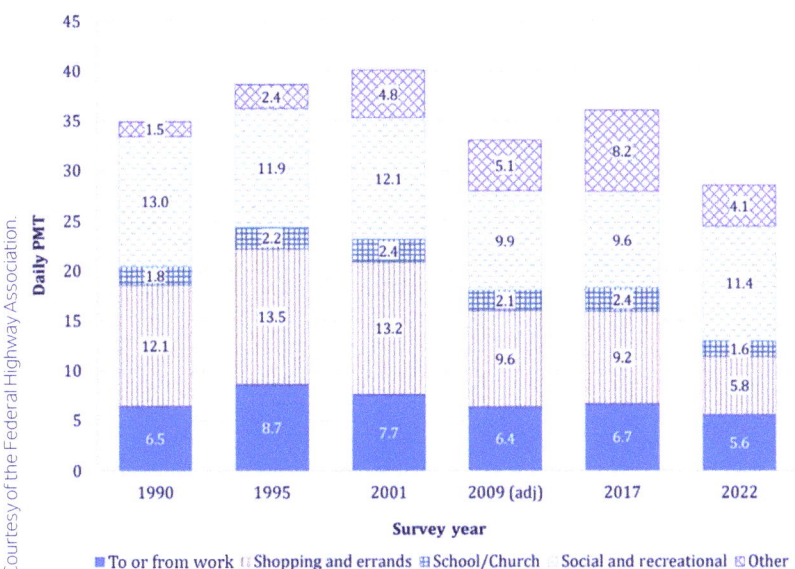

In conclusion, a *logical* vehicle for operating in a city center could have 50 miles range and 30 mph top speed, and for driving over the whole metropolitan area, it may need to have 150 miles range and 70 mph top speed. In either case, it would be a two-seater and shared (because it is used for only one hour a day)!

2.1.1.
Maslow's Needs Hierarchy—Applied to Cars

To understand more about "overengineering" it can be useful to consider Maslow's hierarchy of needs. Developed as a motivational theory in psychology, Abraham Maslow suggests that humans have a hierarchy of needs that must be satisfied in a certain order for individuals to feel motivated and fulfilled. Although Maslow's theory was primarily developed to understand human motivation and behavior, it can also be applied in the context of automobiles and how they cater to different levels of needs.

The most basic functions of an automobile, for example, can be compared with a body's anatomy and physiology. An energy source and propulsion system must provide power and motion, braking and steering systems are required for stopping and changing direction, and a structure is needed to accommodate the occupant(s). This describes a simple vehicle, such as a motorcycle, and is more commonly found in low-income areas (Section 2.2.2).

At the next level up, safety and security needs should be met. This can include seat belts and airbags, lighting systems to illuminate the environment at night, windshield wipers, glazing and doors for protection from rain, and so on.

The third level can be equated with comfort and convenience. This can include heating and air conditioning, audio and navigation systems, comfortable seats, suspension, and good ergonomics. Abundant storage space can be included in this level as well. In a sense, this is where a practical automobile for everyone could stop, with a price well under $20,000 in the US and likely under $10,000 in less developed regions of the world. In essence, the most fundamental requirements for a vehicle are that it should be affordable, reliable, and useful!

Beyond this level, we are into "self-esteem needs," where the owner uses the car to portray an image of status and achievement because they want to feel good and impress others. This is often achieved through performance and styling enhancements, such as additional horsepower, torque and acceleration, and premium materials. This is, effectively, traditional luxury car territory although these enhancements can also be added as aftermarket accessories.

The highest level of Maslow's hierarchy is called "self-actualization" and represents a need for self-fulfillment and seeking meaning in life (**Figure 2.2**). This can manifest itself in eco-friendly materials, environmental propulsion, and cutting-edge infotainment, autonomy, and connectivity solutions that can appeal to a consumer's need for personal growth and new experiences. The content added to enable the top two levels of the pyramid can easily lead to vehicles being able to command a price premium of more than $100,000!

Figure 2.2 Maslow's hierarchy of needs—applied to a car.

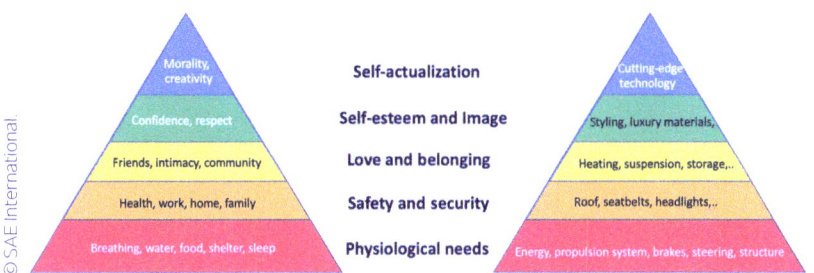

2.1.2.
Automobiles are "Overengineered"

Clearly, even the most basic automobiles are currently being engineered to produce tremendous inefficiencies in time, space, and energy utilization, and the overengineering continues to rise inexorably. From 1990 to 2021, the average weight of a new passenger car increased by 12%, while the average weight of a new pickup truck increased by 32% [2.3]. Because sport utility vehicles (SUVs), vans, and pickup trucks now account for around three in every four new vehicles sold in the US, the average weight of new vehicles has actually increased by 25% over the last 30 years [2.3] and is approximately 30% heavier than in developed Europe and East Asia (**Figure 2.3**). However, the shift towards heavier vehicles is global because, according to the International Energy Agency (IEA), nearly half of all vehicle sales worldwide in 2023 were utility vehicles (either SUVs, built off a truck platform, or crossover vehicles, built off a car platform).

This shift in the US, ironically, may have been stimulated by well-intentioned Corporate Average Fuel Economy (CAFE) standards, which were introduced to improve fuel economy and reduce emissions from vehicles. However, an unintended consequence of these standards was the shift toward larger vehicles (more on this *theme* in Chapter 5), such as SUVs and pickup trucks. This is because larger vehicles, which have a larger footprint, are allowed to meet less stringent fuel economy

Figure 2.3 US auto production share and CO_2 emissions by vehicle segment [2.3].

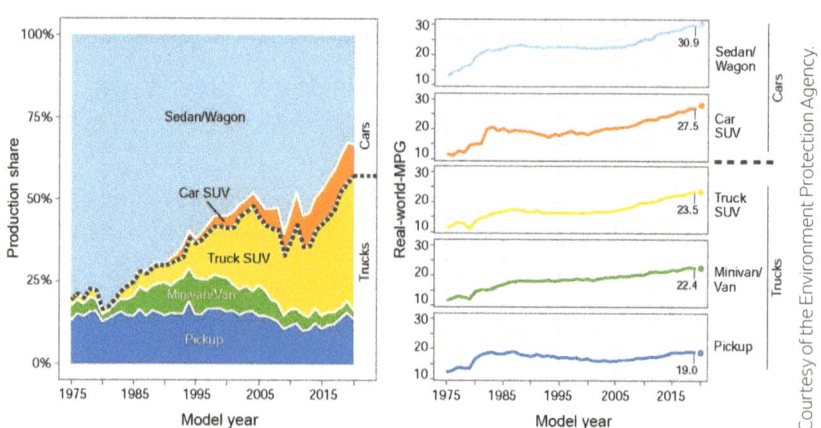

requirements than smaller vehicles. Moreover, a separate government policy allows tax deductions for vehicles that are used for business, and this has encouraged the purchase of more pickup trucks, in particular. In 2021, the average fuel economy of a new truck (including SUVs and vans) was 23.0 mpg, whereas it was 31.8 mpg for a typical new car (including wagons and crossovers) [2.3].

A "vicious circle" can start whereby the growing number of large vehicles on the road may persuade new buyers to buy them because they do not feel as safe on the road in a smaller, lower car when many of the surrounding vehicles are taller, bigger, and heavier. This dynamic can accelerate the adoption of larger vehicles and then be used as "evidence" by automakers that they need to make large vehicles due to customer demand. Lower sales volumes also create more challenging economics for cars, leaving manufacturers little incentive to develop new entry-level, mass-market affordable cars.

The shift toward larger vehicle types has taken place even inside each vehicle segment (**Figure 2.4**). The difference in width, for example, between a first-generation VW Golf (introduced in 1974) and a seventh-generation version (2012) is 181 mm (just over 7 in.). It can be argued that this has been driven, to some extent, by the increased width of the passengers' waistlines over a similar period. For example, the front seat dimensions are driven by a need to accommodate a 95th percentile US male, and waistlines grew from 118.6 cm (46.7 in.) to 125.8 cm (49.5 in.) between two studies that took place just 10 years apart (1988–1994 and 1999–2000) [2.14]. Increased storage and structural reinforcement in the doors, as well as a wider center console, have also caused vehicle width to increase. Modern full-size pickup trucks, in particular, are so wide that the side view mirrors often may need to be powered so that they can retract in order to fit inside a home's garage that may have been built decades ago, and parking spot lines often need to be repainted in recognition of increased vehicle size!

However, vehicle bulging has occurred in all dimensions, not just width. The 1974 Ford F-150 was 205 in. long while the 2024 version is 232 in. long, and even the current Ford Ranger, a compact pickup truck, is 211 in. long. The original 1959 Mini was just 10 ft (3.05 m) long, but the modern version, launched in 2011, is 12 ft (3.7 m) long [2.12]. This increasing size has also come with a mass increase of more than 200 kg, which leads to more energy consumption and pollution during both the production and usage phases. It also makes the vehicles less safe for smaller vehicles, pedestrians, and cyclists in an impact and requires more parking space availability.

Some political actions, particularly in European countries, are challenging this inexorable rise in vehicle size. For example, Denmark and Portugal tax new car registrations progressively based on a vehicle's price and/or engine size, which makes buying larger vehicles significantly more expensive. At the city level, Brussels imposes an annual circulation tax on vehicles registered in the city that increases with engine size, and Paris has higher parking fees for larger vehicles. Urban transport policy relating to cars will be discussed further in Chapter 9, Section 9.5.

Figure 2.4 The width of the average vehicle is creeping up two inches every decade!

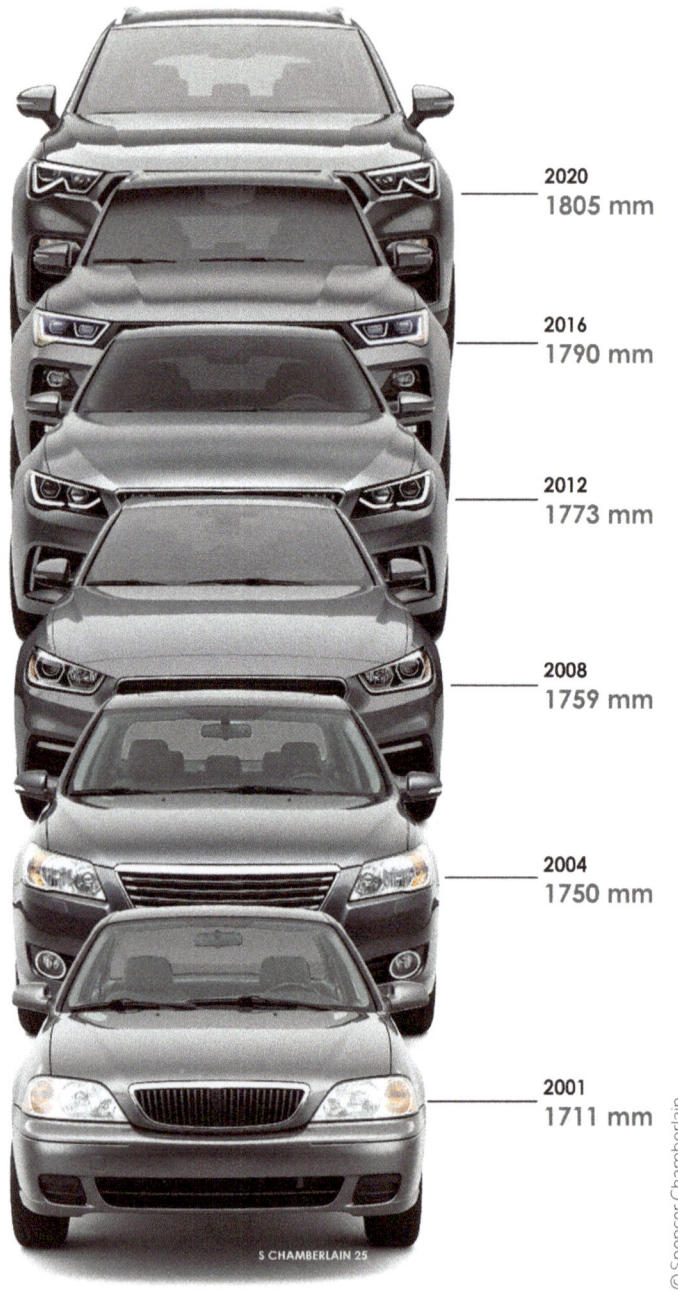

Increasing vehicle size can only partially be explained by the increased size of passengers (see obesity trends later in this chapter) and has much to do with increased vehicle content for safety and comfort reasons. The use of lightweight materials has been essential to keeping vehicle weight from ballooning even more because of this increased vehicle content and, more recently, from the extra weight of batteries in EVs. Since the birth of the motorcar, the materials from which a car has been made have continually evolved. Early cars were often made of wood and heavy steel, and we may yet see a return to biomaterials like wood in the future, as discussed in Chapter 9, Section 9.3. Improvements in vehicle performance (e.g., acceleration, ride and handling, etc.) and a need to meet stringent safety and fuel economy regulations have driven a need for a careful mix of lighter materials, such as aluminum, plastics, composites, and high-strength steels that must balance cost, manufacturability, and performance (strength, durability, etc.).

If lightweight materials had not begun to displace steel, it is likely that fuel economy might have stayed constant since 1990 instead of improving. Recent concerns for battery materials supply chain risk have added another reason for lightweighting as it can help to reduce the amount of battery materials needed. Even so, vehicles have still become slightly heavier over the last 30 years because the expectation for increased safety, comfort, and infotainment content has more than offset the body structure's lighter mass.

However, because of more efficient engines (such as turbocharging) and transmissions (more "gears"), new vehicle fuel economy has actually improved since 2000, even with vehicles becoming heavier. Of course, it would have improved much more if horsepower had actually tracked vehicle weight during this period (**Figure 2.5**) instead of far outpacing it. More powerful engines have been developed to the point where a 2022 Ford Lightning electric pickup truck has a similar 4.5 sec rate of acceleration (for 0–60 mph) as a 2000 Ferrari Modena 360 sports car despite being around twice as heavy, although some of this is due to electric motors having inherently higher torque than ICEs at lower speeds. Simply put, engine development has emphasized power more than efficiency. If vehicles "had not been put on a diet," then engines would have needed to be even bigger to achieve similar acceleration times, and this would have led to heavier vehicles weight because the structure and brake systems would have also been beefed up to support the heavier engine and vehicle!

Figure 2.5 Vehicle performance and size over time [2.3].

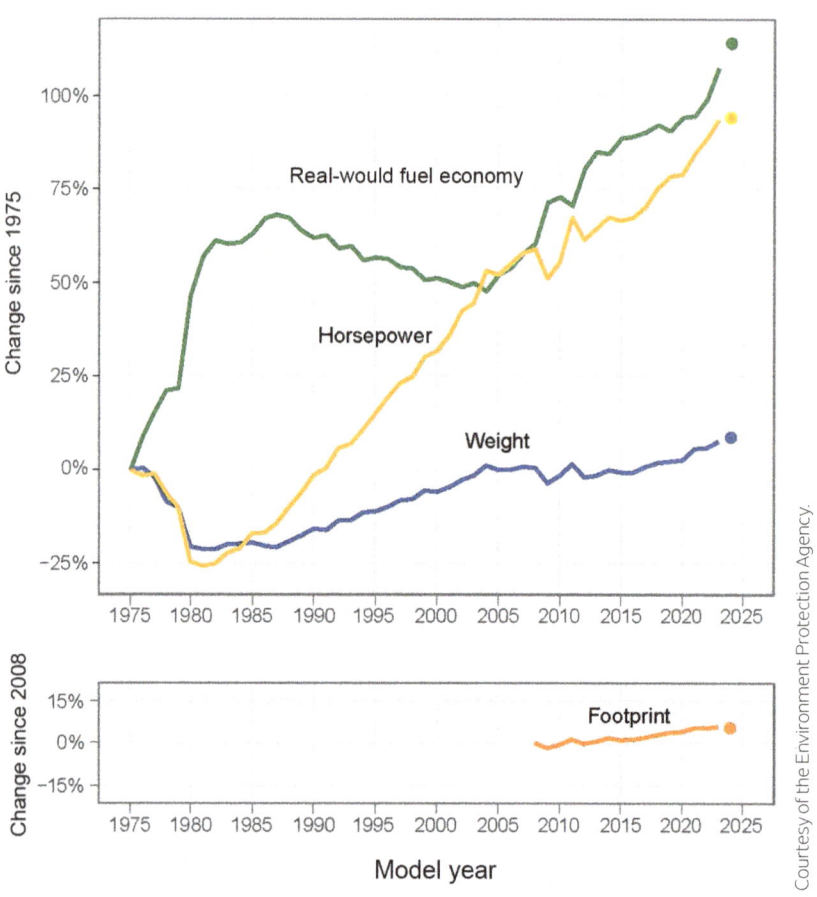

2.2.
The Rise of Automotive Electronics

A vehicle's changing materials composition also reveals that the amounts of copper and silicon, bywords for electrical wiring and electronics, used in a vehicle have tripled in the last 30 years and would be even higher if the comparison was made with a current EV, because some EVs have around 150 lb of copper, which is more than twice as much as a gasoline internal combustion engine vehicle (ICEV) [2.15]. If one word best describes how the car has changed over the last 50 years, then "electronics" is a strong candidate. Without the continuously increasing electronics content in the car, it would be hard to imagine meeting safety, fuel economy, and emissions standards while also offering dramatic improvements in the comfort and convenience features consumers now take for granted, such as the remote key fob, adjustable side mirrors, and navigation displays.

The history of automotive electronics traces its origin back to 1912 when GM introduced the electric starter battery. Ironically, this helped kill off the EV, which was still fighting a battle for supremacy with ICEVs at the time. Although EVs had limited range, they were viewed as less messy (no liquids) and easier to start than the ICEV (no cranking, which was a dangerous procedure as it could cause a broken arm or even jaw). The electric starter battery eliminated the need for hand cranking or "turning over the engine" and sounded the death knell for EVs for the next 100 years!

At around the same time, headlights and then taillights and brake lights were introduced, replacing kerosene lamps. It would take another 20 years for the next automotive electrical innovation, amplitude modulation (AM) radio, to emerge. Radios were initially offered as an aftermarket option and only later became factory-installed equipment, for $130, at a time when an automobile might cost $700 [2.16]. Foreshadowing future developments with new infotainment systems, in a poll conducted by the Auto Club of New York in 1934 [2.17], 56% of respondents thought that the car radio was a "dangerous distraction"! Electrical gauges and instruments, such as speedometers, fuel gauges, and temperature gauges, also gradually became more common in vehicle dashboards during the 1930s.

After World War II, the alternator replaced the direct-current generator, providing a more efficient way to charge the battery and to power the auxiliaries even when idling at a traffic light. Flashing turn signals were introduced to improve visibility and vehicle safety, as did the electric windshield wipers that replaced manual systems (which were operated from inside the vehicle with a lever controlling a spring-loaded arm). At this time, electrical and electronic (E/E) systems (including wiring) accounted for only 1% of the vehicle's content by cost.

In the 1950s, a blind engineer, Ralph Teetor, invented a technology that we now think of as the first step toward vehicle automation because he was not happy with his chauffeur's way of accelerating and braking, and wanted a speed-smoothing solution that could help improve safety, fuel economy, and comfort. It was introduced by Chrysler in 1958 with the tradename "AutoPilot" and did not rely on electronics at all, although it did use electromagnets. A year later when Cadillac sold it with the name "Cruise Control," sales took off and that name stuck.

The next leap in automotive electronics came in the early 1970s with government regulations for meeting more stringent efficiency and emissions standards in the wake of the Clean Air Acts and the Oil Crisis. It was clear that the existing mechanical fuel carburetor could not provide fine enough control of the fuel quantity needed to be injected into the engine, and this led to incomplete combustion, wasted fuel, and tailpipe emissions. The introduction of microprocessors and electronic control units (ECUs) would be applied to the engine systems for fuel injection, ignition timing, and emissions control. In parallel, electronic displays and warning systems for vehicle diagnostics soon came to be visible on the instrument cluster to encourage timely maintenance and to provide the technician with a quick way to assess the problem, leading to on-board diagnostics (OBD) systems that were mandated to monitor and report on the vehicle's emissions control systems. During the 1970s, E/E content typically accounted for around 5% of the vehicle's cost [2.18].

Other vehicle systems became increasingly "electronified" with electronic control replacing the existing mechanical or hydraulic methods that were less precise. Early examples include antilock braking system (ABS) and traction control system (TCS) for improving vehicle stability and active safety, or crash avoidance. Deployment of airbag control modules and advanced restraint systems for enhanced occupant protection was mandated, as was electronic stability control (ESC) to help prevent skids and rollovers. Infotainment systems also began replacing the typical radio and cassette player during the 1980s. In-car CD players, navigation systems, and eventually touchscreen displays became increasingly sophisticated and began to integrate audio, navigation, and communication features. By the mid-1990s, E/E content was climbing to 15–20% of an average vehicle's total cost [2.18], and premium vehicles with a greater number of ECUs had an even higher fraction of E/E content.

Since around 2000, there has been huge growth in electronics content. Advanced driver assistance system (ADAS) features like blind-spot monitoring, lane departure warning (LDW), and adaptive cruise control (ACC) are becoming more common because of the integration of advanced sensors, such as radars and cameras, that can "see" the surroundings and either alert the driver or communicate with the brakes directly to take over if the driver is not able to respond quickly enough, in the case of automatic emergency braking (AEB). At the same time, advanced human–machine interfaces (HMI), such as voice recognition, gesture control, and augmented reality displays, are beginning to reshape how the driver and the other occupants communicate inside the vehicle and with the outside world. As vehicles become connected to the Internet, enabling remote diagnostics, software updates, and infotainment services, there is now real-time data exchange between vehicles, infrastructure, and cloud services. This enables over-the-air (OTA) software updates to remotely improve vehicle systems, but it also means we need more robust cybersecurity methods to protect against potential hacking and unauthorized access to all vehicle systems.

Modern vehicles are essentially "computers or smartphones on wheels," often with 100 or more ECUs managing various systems and an increasing number of ADAS features and connectivity options. The integration of electronics in automobiles continues to evolve rapidly, driven by consumer demand for safety, convenience, and efficient mobility to a point where E/E content is now responsible for around 30% of the entire vehicle's cost [2.19], and this number is expected to grow, reinforced by the simultaneous growth of electric propulsion.

2.2.1.
The Life of a Typical Car—a Link between Developed and Developing Worlds

So far, the discussion has been around the automobile's design and ownership in the developed world, with most of the data coming from US sources. Even though vehicles tend to be somewhat smaller and are driven less in other developed world countries, and there is often an effective public transport alternative, the conclusions around overengineering (and of future demographic trends) still generally apply.

The vehicle fleet tends to turn over roughly every 15 years, consistent with a typical automobile having an 18-year lifespan from production date to scrappage [2.20, 2.21]. This length of time is a major impediment to the adoption of any new technology and, in the case of EVs, it means that the maximum environmental benefits of a changing vehicle parc are delayed by a decade or so. A new vehicle will likely be under warranty for up to five years [2.22]. But after it reaches about 100,000 miles, an entry-level or mid-level vehicle may be exported to the developing world [2.23], where the demand for affordable mobility is high (**Figure 2.6**).

Figure 2.6 Current regulatory regime for importing used vehicles [2.24].

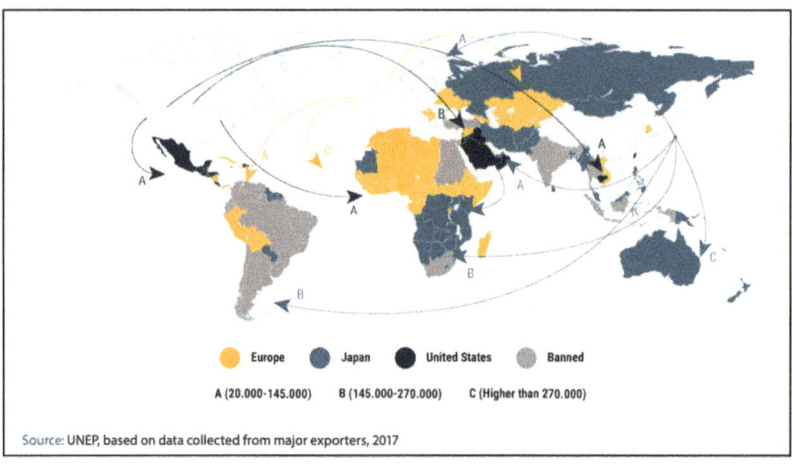

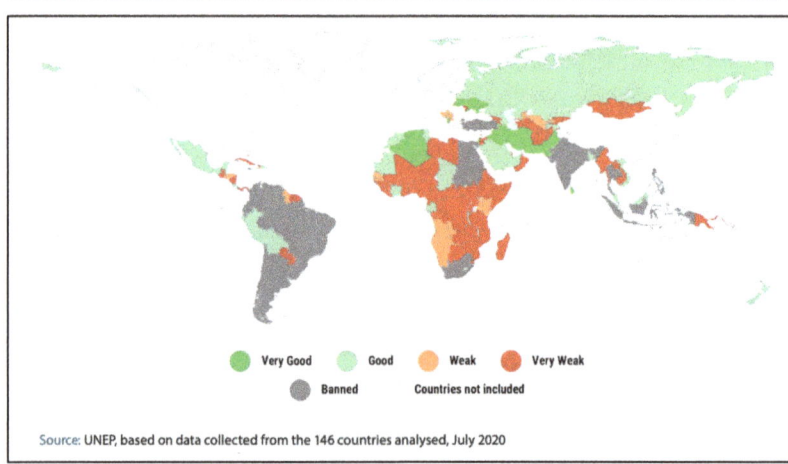

The United Nations (UN) estimates that 14 million vehicles were exported to the developing world between 2015 and 2018, with Africa being a destination for 40% of them [2.24]. Depending on the buyer, the car might undergo pre-export inspection and refurbishment to meet the importing country's regulations (e.g., steering wheel location, emissions standards), but some of these imported vehicles may be heavily emitting, which raises concerns that they are being "dumped" [2.24]. Eventually, the car may be dismantled for parts or scrapped entirely, depending on local regulations and dismantling infrastructure. Valuable parts will be retrieved for reuse or resale and the scrap metal is recycled. In some cases, the car might be exported to developing countries specifically for dismantling and parts resale, but this practice has raised environmental concerns because of lax regulations in some countries. Concerns for air pollution and potential negative health impacts are leading some African countries to consider stricter regulations on used car imports [2.25].

Just 2% of all new vehicles sold each year globally are in African countries, and approximately 90% of vehicles imported into Africa are used vehicles from other continents [2.26]. The preferred method of vehicle production in many parts of the developing world is complete or semi-complete knockdown (CKD) assembly, whereby vehicles are assembled locally using parts imported from the original automaker [2.27] because this is significantly less costly than building a new plant for low-volume production. For the host country, it can be the first step toward creating a trained workforce for future manufacturing of automobiles and other products.

Despite importing used cars and having knockdown assembly plants, there are only approximately 50 automobiles per 1000 population in sub-Saharan African countries and around 200 in Southeast Asia and Latin America. This compares with more than 600 vehicles per 1000 population in North America and Western Europe [2.28]. Clearly, more affordable solutions are required for the vast majority of the world's population. OX is developing a rugged electric cargo van that can be recharged from a solar microgrid and has an approximate payload capacity of 4000 lb. It is specifically designed for rugged terrain, having a high ground clearance, large overhangs, and steep approach and departure angles (**Figure 2.7**), and is being exported to East Africa as a flat pack CKD kit that is easy to assemble and has relatively few parts. It also comes with a novel financing strategy, whereby customers, who are typically farmers, only pay for the portion of the 8.8 cubic meter cargo space that they need. The weight per distance traveled is claimed to be comparable to a motorbike but with higher capacity and speed [2.29].

Figure 2.7 OX vehicle. OX Delivers a new type of truck and business model to East Africa.

2.2.2. Other Motorized Passenger Vehicles

In addition to affordability, another barrier to automobile adoption in parts of the developing world is the state of the roads, which may be poorly maintained and can be very rough, pitted with large potholes and cracks, and subject to flooding in the rainy season. In addition, many developing countries do not have well-connected road networks, especially in rural areas, and often lack signage, traffic signals, and illumination. Rural communities often lack access to paved roads, and navigating these roads and avoiding a tire puncture is far easier with a motorcycle than with a car. In dense, urban centers, motorcycles can also overcome the traffic congestion that snarls up car travel. For all these reasons, two-wheelers are the most popular choice for long-distance personal transport in most parts of the developed world.

Globally, there are approximately 1.3 billion automobiles on the roads worldwide and 600 million motorcycles [2.31, 2.32], but the two vehicle types are distributed very differently. Nearly all the motorbike sales are in the developing world, concentrated in South and East Asia, and outnumber car sales by more than 5:1 in India. The relative popularity of two-wheelers (2W) in India is shown in **Figure 2.8**.

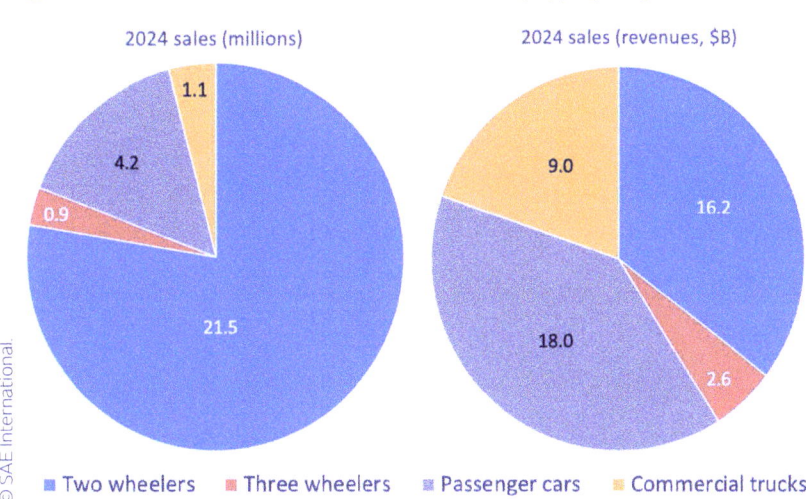

Figure 2.8 2024 Sales of motor vehicles in India by type [2.30].

According to data from the Asian Development Bank, motorbikes make up more than 80% of all vehicles in several Southeast Asian nations like Vietnam, Thailand, Malaysia, and Indonesia. Government registration data show that Vietnam had more than 58 million motorbikes as of 2020 but only around 3 million cars (in 2018) [2.33]. In 2019, Thailand had around 22 million motorbikes and 10 million cars [2.34]. And in 2018, Indonesia had 125 million registered motorbikes versus 17 million registered cars [2.35]. Car ownership is likely to increase over time as countries develop economically, but it is unlikely to overtake motorbike ownership anytime soon.

In 2023, global sales of motorbikes reached 62.5 million, compared with 76.7 million automobiles in the same year [2.36, 2.37]. Global motorcycle revenue is around $110 billion [2.3] versus around $2.5 trillion for the automotive industry, indicating that the average motorcycle costs under $2000 while the average car globally costs over $30,000.

Tuk-tuks (or auto-rickshaws) are covered, gasoline-powered tricycles and are often used to provide a taxi service in cities inside the Indian subcontinent and Southeast Asia, but they can also be used to deliver goods. Precise global sales data for tuk-tuks are difficult to find, but BloombergNEF estimates the global fleet size for three-wheeled vehicles (3W) is around 120 million [2.38]. EMBARQ India's analysis, based on government records, indicates that Tier I cities in India (with a population exceeding 4 million) tend to have around 50,000 auto-rickshaws, while Tier II cities (with a population between 1 and 4 million) have 15–30,000 auto-rickshaws. This is a significant share of the vehicle mix in cities, and as a rough rule of thumb, there may be about ten auto-rickshaws per 1000 people in Tier I and II cities [2.39]. As will be discussed in Chapter 3, Section 3.1 electric drive is rapidly gaining 3W market share from gasoline engines.

Motorbikes and 2W are also very popular across Africa and greatly outnumber automobiles. Estimates suggest that there are approximately 8 million motorbikes in Nigeria [2.40], Africa's largest country by population, and 27 million registered motorbikes across sub-Saharan Africa in 2022 [2.41], versus below 5 million as recently as 2010, according to the FIA Foundation [2.42]. However, it should be noted that many sub-Saharan African countries lack comprehensive data collection on vehicle usage, especially for informal motorbike travel in rural areas. Because the road quality and availability differ greatly across the region, it is likely that the more developed urban areas have relatively more car usage than rural areas but still less than motorcycle usage because the latter can handle road congestion more easily and can provide a less expensive and faster taxi service. Due to affordability, fuel efficiency, and navigating unpaved roads, the average mileage driven on motorbikes is probably significantly higher than for cars in sub-Saharan Africa [2.43].

In Latin America, motorbike usage is lower compared to Africa or Southeast Asia. A 2018 report by Agência Brasil indicates that motorcycles outnumbered cars in nearly half of the cities, with an estimated 26.4 million motorcycles nationwide, but cars remain the leading mode of transportation overall because the same Agência Brasil report estimates that there are 53.4 million cars in Brazil [2.44].

For regions of the developing world that have navigable rivers, boat travel can also play a crucially important role in moving people and goods in some areas. Boats are essential for specific regions and short-distance commutes in areas with canals and rivers. They are also used in the archipelagos that are part of Indonesia, as well as for tourist cruises, such as along the Mekong River in Vietnam. Much boat travel, especially on the smaller waterways, might be informal and undocumented, making data collection challenging.

Although not expected to be used for moving people in forested areas, drones can play a role in transporting goods, such as medical supplies, aerially to communities that may be hard or slow to reach by road and will be discussed in later chapters.

2.3.
Future Societal and Demographic Trends

In addition to technology advances, societal trends are another force driving the future of vehicle development. These trends include urbanization, demographics, and obesity.

2.3.1.
Urbanization

Urbanization, for example, is a major trend affecting future personal mobility because 68% of the world's population is projected to live in urban areas by 2050, up from 55% today, according to the UN [2.45]. Already in the developed world, approximately 80% of people live in metropolitan areas but this percentage is unlikely to grow much further (**Figure 2.9**).

Figure 2.9 Share of the population living in urban areas [2.46].

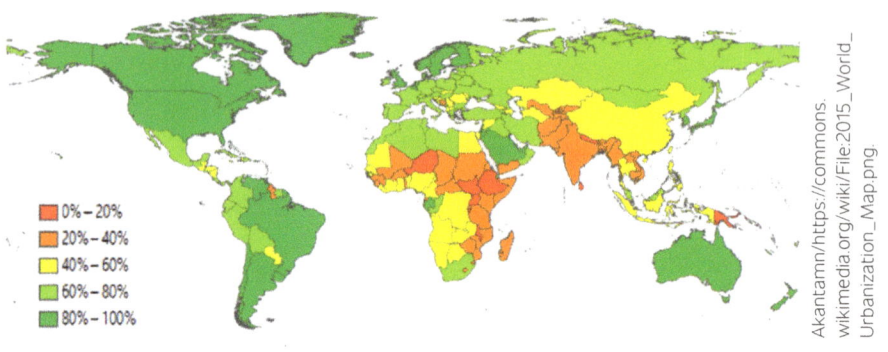

However, in Asia and Africa, the percentage of people living in urban areas has only recently begun to approach 50% and is expected to grow at a significant rate for the foreseeable future. There are around 500 cities worldwide with a population of at least one million, and 44 of them are megacities that have a population of more than 10 million. The number of megacities is projected to grow to 67 by 2050 [2.47], and nearly all of them are in the developing world.

Much of this migration is to densely populated cities that promise economic opportunity but also tend to have congested traffic, limited parking, and poor air quality. As more people move to big cities, it places greater emphasis on vehicles to be ultracompact for easier parking and maneuvering, and to have low or even zero tailpipe emissions. It also decreases the likelihood of them being used at all because automobile ownership tends to decrease as population density increases. If the population density is high enough, as can be the case in many cities, even affluent people may choose not to own a car (**Figure 2.10**).

Figure 2.10 Population density, household income and vehicle ownership, using New York and its five boroughs as an example [2.48].

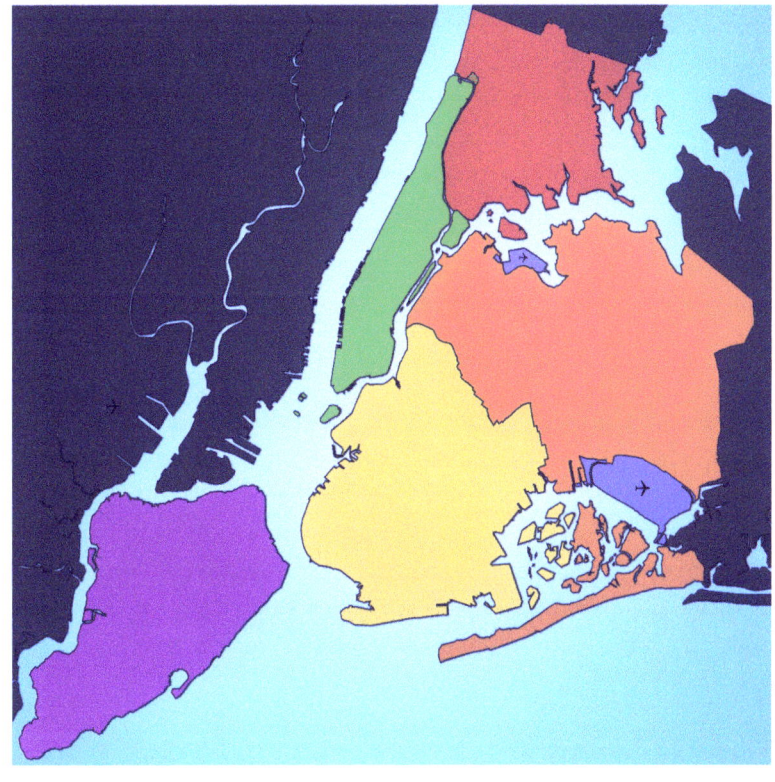

	Population density (per square mile)	Median household income ($)	Percent of households owning at least one car
Manhattan	72,918	85,000	23
Brooklyn	38,634	62,000	44
Bronx	34,920	41,000	40
Queen's	21,460	74,000	65
Staten Island	8,618	82,000	84

As **Figure 2.10** shows, there tends to be an inverse relationship between population density and vehicle ownership. In the case of New York City, the exception to this is between Bronx and Brooklyn and may be due to level of income. For example, in New York City's densely populated (and affluent) Manhattan borough, the vast majority of residents do not own a vehicle. However, as one radiates away from Manhattan to the other boroughs, vehicle ownership increases and is particularly high in suburban Staten Island, where nearly all households have at least one vehicle [2.49]. In addition to the deterrents of finding and paying for parking, and slow-moving traffic, Manhattan has convenient ride-hailing and an extensive subway system that offers an attractive alternative to vehicle ownership. Most major cities outside the US are also densely populated, because there is less urban sprawl than in American cities like Los Angeles and Houston [2.50]. For example, New York City's population density is approximately 27,000 people per square mile, and the corresponding numbers for Shanghai, Mumbai, Mexico City, Sao Paolo, Cairo, Lagos, Moscow and London, to name a few large cities outside the US, range between 8500 (Moscow) and 73,000 (Mumbai).

In most parts of the world, automobile owners tend to benefit from subsidized city streets and store parking (where the costs are paid for in general taxes and rents) [2.51]. This has been criticized by some for skewing economic policy in favor of vehicle usage because these costs are passed onto the taxpayer or local residents instead of directly to the vehicle users. In some cases, governments promote vehicle production, because it creates high-paying jobs and is a driver for economic development. On the other hand, they often wish to deter vehicle usage. This is accomplished through methods such as a congestion charge (Singapore) [2.52], or license plate restrictions on which day of the week the vehicle cannot be used (quite common in some Chinese and Indian cities).

Public transport (e.g., trains, buses, bus rapid transit, trams) plays an increasingly vital role in moving people efficiently in crowded metropolitan areas [2.53], and as space efficiency becomes paramount, micromobility solutions with a smaller footprint, like bicycles and electric scooters, become viable options because they can decrease travel times door-to-door in the city center and be very cost-effective [2.54]. This will be covered more in Chapter 7, Section 7.2.

2.3.2.
Age Demographics

Figure 2.11 Younger people in the US are driving less than previous generations [2.55].

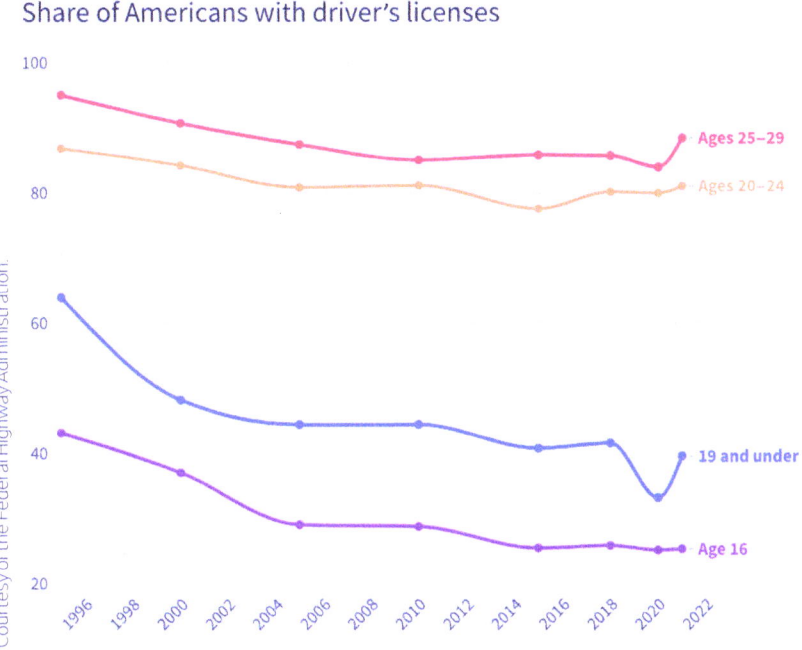

Somewhat related to the urbanization trend is the shifting preferences among youth away from vehicle ownership. In surveys across the US and Europe, it seems that younger people are less interested in obtaining a driver's license and purchasing automobiles compared to their parents' generation (**Figure 2.11**). For example, in 2022 only approximately 40% of people under 20 had a driver's license, compared to around 65% of people in the same age group in 1996, according to the Federal Highway Administration [2.55]. As a trend, wealthier youths increasingly prefer on-demand access and transport as a service over personal ownership with its attendant responsibilities for financing, insurance, maintenance, refueling, and parking. A consumer survey, conducted by Deloitte, on the future of automotive mobility, found that the recent rise of on-demand mobility may have caused half of

European youths and a third of US youths to question the need to own a car [2.56]. This is particularly the case for youths who live in dense, urban centers that have viable alternatives to vehicle ownership. For many others, the cost of buying a car, insuring it, and learning to drive can be a challenge, particularly in lower-income families. Millennials drive less even when socioeconomic differences are taken into account because higher use of smartphones and the Internet (such as social media and online shopping) also plays a role. This means that future mobility frameworks must integrate micromobility solutions, like electric scooters and e-bikes, with seamless access to public transport and automobile access when needed [2.56]. Fortunately, smartphones with GPS, mobile payments, and Internet connectivity can enable such a transformative approach and will be covered extensively in Chapter 7.

At the other end of the age spectrum, global life expectancy increased by nearly seven years between 1990 and 2019, according to the World Health Organization (WHO) [2.57], and in developed nations over 20% of the population is already aged 65 or older [2.58]. As this elderly demographic continues growing, mobility solutions must adapt to age-related impairments like declining vision and cognition, slower reaction times, and restricted mobility that affect safe vehicle operation. Designing for the elderly is often considered a "universal design" practice because accommodations for the elderly tend to benefit everyone [2.59]. Specific automotive considerations include making it easier to get in and out of the vehicle, making it easier to see other road users when driving, designing ergonomic controls, and ultimately providing fully AV operation. However, even in the case of full AVs with no opportunity for manual driving, there will still be a need to make it easy to enter, exit, and "direct" the vehicle to where it needs to go, and dedicated mobility services and transportation alternatives may still be required for those unable to personally operate a vehicle due to advanced age or disabilities.

Will an aging population result in fewer new vehicles being sold? To a large extent, that depends on the health and activity of the elderly and where they choose to live, which affects the availability of alternatives to vehicle ownership, such as public transport, ride-hailing, and, potentially, robotaxis. However, it is reasonable to expect the need for in-vehicle health and wellness monitoring, online goods delivery, and multimodal vehicle integration to increase. This will be covered in later chapters.

2.3.3. Obesity

Another trend is increasing obesity, which has nearly tripled worldwide since 1975. Nearly 900 million people are considered obese, and another 1.6 billion adults are considered overweight, according to the WHO [2.60].

Obesity rates, not surprisingly, are highest in the developed world, where there are high rates of car ownership (**Figure 2.12**). This is probably due to a combination that includes reduced physical exercise and increased caloric intake. Obesity has some of the same vehicle design implications as the aging trend because accessibility is a common challenge. Excess weight increases the likelihood of chronic immobility and will require personal mobility solutions that also facilitate entry/exit and accommodate larger body types, as hinted at earlier in this chapter. This will likely mean larger and wider door openings, swivel seating, and reinforced structural support, as well as easily accessible and operable controls for the vehicle. In-vehicle health and wellness monitoring will also be increasingly useful and could become a common feature in some new vehicles by 2035, perhaps combined with wearables. In extreme cases, being mobile may require an accessibility service and/or AV operation.

Figure 2.12 Global obesity rates [2.61].

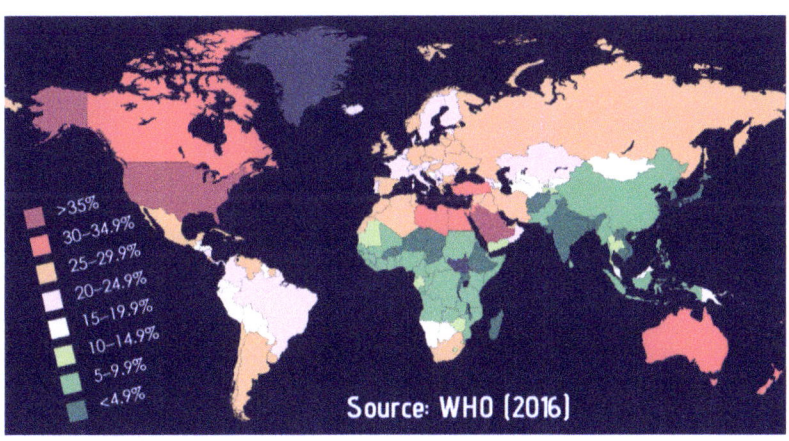

Some Observations from Sub-Saharan Africa

A "Gas Station" in rural Mali.

Back in February 2009 and August 2010 I visited a rural Mali village, Nana Keneiba, with my two daughters, Sophia and Christina, as part of a research project led by Professor Mark Bryden (at Iowa State University). The village had a population of around 700 and was approximately 50 miles from the capital city (Bamako). The most likely motorized vehicle one might have seen was a motorcycle or minibus. Building a gasoline station is not practical in such a location, but people still need access to gasoline, so an entrepreneur in the village decided to pour gasoline into old wine bottles and sell them individually to customers.

In Ghana's capital, Accra, in March 2017 I visited an auto parts market with Sophia, who was studying there on a Fulbright Scholarship. I was intrigued by the adjacent market stalls that were dedicated to nearly all vehicle components, with a large, random selection of parts from nearly every car model. These parts had probably been harvested from scrapped vehicles at the end of their lives.

Courtesy of Chris Borroni-Bird.

Auto parts market in Accra, Ghana (e.g., engine blocks, exterior lights, fluid reservoirs, lead acid batteries).

Metals recovery from the Agbogbloshie e-waste site in Accra, Ghana.

During the same trip and across town, I was deterred from visiting Agbogbloshie, an e-waste site where recyclers burn, dismantle, and sell old electronics imported from richer countries, because the resulting air pollution is so toxic. Recyclers burn plastic lead acid battery containers to remove valuable lead metal, but this releases harmful pollutants, such as particulates, heavy metals, and toxic chemicals, that far exceed the air quality standards recommended by the WHO. This air pollution can cause respiratory problems, heart disease, and cancer, and can affect male and female fertility. There are efforts to improve the situation, but health risks remain significant for the workers.

While in Zimbabwe in March 2020 with my colleague Richard Saad, I noticed long lines at the gas station. As soon as there was notification of a gasoline truck making its way to the refueling station, people would begin queuing, and the line for refueling lasted for days. People would sleep in their car, and if possible, a relative or friend would switch places with the driver to give them a break from being in the car. I should point out that this situation is not typical and is not limited to sub-Saharan Africa—it also happened in other countries, such as Sri Lanka in 2022.

References

2.1. Center for Sustainable Systems, University of Michigan, "Personal Transportation Factsheet," Pub. No. CSS01-07, 2023, accessed June 2024, https://css.umich.edu/sites/default/files/2023-10/Personal%20Transportation_CSS01-07.pdf.

2.2. United States Census Bureau, "B08301 Means of Transportation to Work," American Community Survey, accessed June 2024, https://data.census.gov/cedsci/table?q=B08301&tid=ACSDT1Y2019.B08301.

2.3. EPA, "Greenhouse Gas Emissions, Fuel Economy, and Technology since 1975, Executive Summary," The 2024 EPA Automotive Trends Report, EPA-420-S-24-001, November 2024, https://nepis.epa.gov/Exe/ZyPDF.cgi?Dockey=P101CUZD.pdf.

2.4. What Things Weigh, "How Much Does a Bus Weigh?," accessed June 2024, https://whatthingsweigh.com/how-much-does-a-bus-weigh/.

2.5. Triplett, T., Santos, R., Rosenbloom, S., and Tefft, B., "American Driving Survey: 2014–2015," AAA Foundation for Traffic Safety, Washington, DC, 2016, https://aaafoundation.org/wp-content/uploads/2017/12/AmericanDrivingSurvey2015.pdf.

2.6. NYC Department of Transportation, "New York City Mobility Report," August 2019, https://www.nyc.gov/html/dot/downloads/pdf/mobility-report-2019-print.pdf.

2.7. Varga, B., Sagoian, A., and Mariasiu, F., "Prediction of Electric Vehicle Range: A Comprehensive Review of Current Issues and Challenges," *Energies* 15 (2019): 946, doi:10.3390/en12050946.

2.8. INRIX, "INRIX Global Traffic Scorecard," accessed June 2024, https://inrix.com/scorecard/.

2.9. https://www.tomtom.com/traffic-index/los-angeles-ca-traffic/.

2.10. U.S. Department of Transportation, "Summary of Travel Trends 2022 National Household Travel Survey," https://nhts.ornl.gov/assets/2022/pub/2022_NHTS_Summary_Travel_Trends.pdf.

2.11. Jolly, J., "'Yes, Lego Car!': Why Small Electric Cars Could Be about to Break the Grip of SUVs," *The Guardian*, June 1, 2024, https://www.theguardian.com/business/article/2024/jun/01/yes-lego-car-why-small-electric-cars-could-be-about-to-break-the-grip-of-suvs.

2.12. UK Department of Transport, "National Travel Survey: England 2019," August 5, 2020, https://assets.publishing.service.gov.uk/media/5f27f7748fa8f57ac683d856/national-travel-survey-2019.pdf.

2.13. Cox, W., "Average Chinese Car Travels as much as an American Car," Newgeography Website, September 23, 2019, https://www.newgeography.com/content/006420-average-chinese-car-travels-much-american-car.

2.14. Ford, E.S. et al., "Trends in Waist Circumference among US Adults," https://doi.org/10.1038/oby.2003.168.

2.15. Nguyen, M., "Innovation in EVs Seen Denting Copper Demand Growth Potential," Reuters, July 9, 2023, https://www.reuters.com/business/autos-transportation/innovation-evs-seen-denting-copper-demand-growth-potential-2023-07-07/.

2.16. "The Fascinating History of Car Radio," Updated October 26, 2023, https://www.carcovers.com/articles/the-fascinating-history-of-the-car-radio.

2.17. Womack, J.P., Jones, D.T., and Roos, D., *The Machine that Changed the World* (1990).

2.18. Jurgen, R.K., *Automotive Electronics Handbook* (New York: McGraw-Hill, 1999).

2.19. "Tomorrow's Automotive Chips," IEEE Spectrum, March 2022.

2.20. Schwartz, H., "America's Aging Vehicles Delay Rate of Fleet Turnover," The Fuse, January 23, 2018, https://soltron-gtr.com/wp-content/uploads/2018/07/America%E2%80%99s-Aging-Vehicles.pdf.

2.21. Held, M. et al., "Lifespans of Passenger Cars in Europe: Empirical Modelling of Fleet Turnover Dynamics," *European Transport Research Review* 13 (2021): 9.

2.22. "Manufacturer Warranty: Length and Coverage Terms (2024)," Updated January 30, 2023, https://www.carchex.com/content/manufacturer-warranty.

2.23. Dror, M.B. and Vahle, T., "How Used Car Exports to Africa Could Become the Development Opportunity of the Decade," World Economic Forum, January 18, 2023, https://www.weforum.org/agenda/2023/01/used-car-exports-to-africa-development-opportunity-davos-2023/.

2.24. Fabian, B., "New UNEP Report Highlights Importance of Used Vehicle Flows for Fuel Economy, Emissions, and Vehicle Safety in Developing Countries," Global Fuel Economy Initiative, November 2, 2020, https://www.globalfueleconomy.org/news/2020/november/new-unep-report-highlights-importance-of-used-vehicle-flows-for-fuel-economy-emissions-and-vehicle-safety-in-developing-countries.

2.25. UNEP Press Release, "New UN Report Details Environmental Impacts of Export of Used Vehicles to Developing World," October 26, 2020, https://www.unep.org/news-and-stories/press-release/new-un-report-details-environmental-impacts-export-used-vehicles.

2.26. Kuhudzai, R.J., "Electric Cars May Come to Several African Countries a Whole Lot Faster than Most People Think," CleanTechnica, July 27, 2024, https://cleantechnica.com/2024/07/27/electric-cars-may-come-to-several-african-countries-a-whole-lot-faster-than-most-people-think/.

2.27. Abate, E. et al., "Identifying Standard SKD/CKD and Automotive Manufacturing Development Stage in Ethiopia," June 2023, https://www.researchgate.net/publication/371959658_Identifying_Standard_SKDCKD_and_Automotive_Manufacturing_Development_Stage_in_Ethiopia.

2.28. OICA, "International Organization of Motor Vehicle Manufactures," https://www.oica.net/category/vehicles-in-use/.

2.29. OX Delivers, "The World's First Purpose-Designed Electric Truck for the Global South," https://www.oxdelivers.com/solution.

2.30. Society of Indian Automobile Manufacturers, "Performance of Indian Auto Industry in 2023-24," https://www.siam.in/statistics.aspx?mpgid=8&pgidtrail=9.

2.31. Bloomberg New Energy Finance, "Electric Vehicle Outlook 2024," https://about.bnef.com/electric-vehicle-outlook/.

2.32. Riders-share.com, "How Many Motorcycles Are There in the World?," February 24, 2023, https://www.riders-share.com/blog/article/number-motorcycles-world-top-countries.

2.33. Huu, D.N. and Ngoc, V.N., "Analysis Study of Current Transportation Status I Vietnam's Urban Traffic and the Transition to Electric Two-Wheelers Mobility," *Sustainability* 13, no. 10 (2021): 5577, doi:https://doi.org/10.3390/su13105577.

2.34. Apisitniran, L., "Produces Maintain Motorcycle Output Target," Bangkok Post, June 4, 2024, https://www.bangkokpost.com/business/motoring/2804734/producers-maintain-motorcycle-output-target.

2.35. Statista Website, "Number of Vehicles in Indonesia from 2018 to 2022, By Type," https://www.statista.com/statistics/1239274/indonesia-number-of-vehicles-by-type/.

2.36. MotorCycles Data, "World's Best Selling Motorcycles Ranking 2023," March 14, 2024, https://www.motorcyclesdata.com/2024/03/14/best-selling-motorcycles/#:~:text=During%20the%202023%2C%20the%20global,reported%20for%20the%20electric%20segment.

2.37. Global Market Insights, "Motorcycle Market – By Type," Report ID: GMI7762, January 2024, https://www.gminsights.com/industry-analysis/motorcycle-market.

2.38. BloombergNEF, "Electric Vehicle Outlook 2024," June 12, 2024, https://about.bnef.com/electric-vehicle-outlook/.

2.39. Mani, A., "Market Size of Auto-rickshaws in Indian Cities," WRI India Ross Center, May 27, 2014, https://www.wricitiesindia.org/content/market-size-auto-rickshaws-indian-cities.

2.40. World Population Review, "Motorcycles by Country 205," https://worldpopulationreview.com/country-rankings/motorcycles-by-country.

2.41. Zote, A., "A Rise in Motorcycles New for Sub-Saharan Africa," ESI Africa, November 21, 2022, https://www.esi-africa.com/news/a-rise-in-motorcycles-new-for-sub-saharan-africa/.

2.42. Bishop, T. and Courtright, T., "The Wheels of Change: Safe and Sustainable Motorcycles in Sub-Saharan Africa," FIA Foundation, 2022, https://motorcycleminds.org/virtuallibrary/strategies/the_wheels_of_change_safe_and_sustainable_motorcycles__in_sub_saharan_africa_nov_2022.pdf.

2.43. Bourdache, J., "Africa: The Motorcyclist's Continent," The Vintagent, March 5, 2018, https://thevintagent.com/2018/03/05/africa-the-motorcyclists-continent/.

2.44. agenciaBrasil, "Motorcycles Outnumber Cars in 45% of Brazilian Cities," Translated by Fabricio Ferreira, September 7, 2018, https://agenciabrasil.ebc.com.br/en/geral/noticia/2018-07/motorcycles-outnumber-cars-45-brazilian-cities.

2.45. United Nations, "Department of Economic and Social Affairs: Population Division," https://population.un.org/wup/.

2.46. Our World in Data, "Share of the Population Living in Urban Areas," https://ourworldindata.org/grapher/share-of-population-urban?tab=chart&country=Low-income+countries~High-income+countries~Middle-income+countries~Lower-middle-income+countries~Upper-middle-income+countries.

2.47. Oxford Economics, "Rise of New Megacities Will Drive Global Urban Growth," January 26, 2024, https://www.oxfordeconomics.com/wp-content/uploads/2024/01/Rise-of-new-megacities-will-drive-global-urban-growth.pdf.

2.48. "2013-2017 5-Year ACS (US Census); Analysis by Trulia," https://cityobservatory.org/2019/05/.

2.49. Tri-State Transportation Center, "How Car-Free Is New York City?," April 21, 2017, https://blog.tstc.org/2017/04/21/car-free-new-york-city/.

2.50. Ritchie, H., Samborska, V., and Roser, M., "Urbanization," Our World in Data, Revised February 2024, https://ourworldindata.org/urbanization.

2.51. Caldwell, N., "How Driving Is Subsidized in America," stacker.com, April 26, 2022, https://stacker.com/society/how-driving-subsidized-america.

2.52. Ramos, R. et al., "From Restricting the Use of Cars by License Plate Numbers to Congestion Charging: Analysis for Medellin, Colombia," *Transport Policy* 60 (2017): 119-130.

2.53. American Public Transportation Association, "Public Transportation Facts," https://www.apta.com/news-publications/public-transportation-facts/.

2.54. Abdelkareem, M.A. et al., "Micromobility: Progress, Benefits, Challenges, Policy and Regulations, Energy Sources and Storage, and Its Role in Achieving Sustainable Development Goals," *International Journal of Thermofluids* 17 (2023): 100292.

2.55. de Vise, D., "American Teens Are Driving Less," The Hill, July 26, 2023, https://thehill.com/policy/transportation/4119244-american-teens-are-driving-less/#:~:text=The%20share%20of%20teenagers%20with,prices%20of%20insurance%20and%20gas.

2.56. Severen, C., "Why Are Young People Driving Less? Evidence Points to Economics, Not Preferences," Brookings, March 24, 2023, https://www.brookings.edu/articles/why-are-young-people-driving-less-evidence-points-to-economics-not-preferences/.

2.57. World Health Organization, "Global Health Estimates: Life Expectancy and Healthy Life Expectancy," Global Health Observatory, https://www.who.int/data/gho/data/themes/mortality-and-global-health-estimates/ghe-life-expectancy-and-healthy-life-expectancy.

2.58. Eurostat, "More than a Fifth of the EU Population Are Aged 65 or Over," March 16, 2021, https://ec.europa.eu/eurostat/web/products-eurostat-news/-/ddn-20210316-1.

2.59. Centre for Excellence in Universal Design, "The 7 Principles," https://universaldesign.ie/about-universal-design/the-7-principles.

2.60. World Health Organization, "Obesity and Overweight," March 1, 2024, https://www.who.int/news-room/fact-sheets/detail/obesity-and-overweight.

2.61. Wisevoter Website, "Obesity Rate by Country," https://wisevoter.com/country-rankings/obesity-rates-by-country/.

Chapter 03

The Recent Automotive (R)evolution 1990–2025

The auto industry is currently being disrupted by three fundamental technological forces: replacement of the ICE powertrain by battery-powered electric motors, automation of the vehicle controls (for braking, steering, and accelerating), and connectivity to and from the vehicle that allows information to be shared with other vehicles and road users, the road infrastructure, and the cloud. In short, the paradigm of ICEVs driven by humans and operating as standalone products on the road is giving way, after more than a century, to EVs that can drive by themselves and communicate back and forth with each other to learn and self-improve. The promise is that this will lead to a cleaner, safer road environment that can improve traffic coordination.

In this chapter, we will cover recent progress in these three areas and how they come together to reinvent the automobile. The year 1990 is a reasonable starting point as, in hindsight, one can detect an increase in activity in these three areas around that time, which has continued to gather momentum ever since. We will start with EVs, which have actually been researched for almost as long as the car has been around.

3.1.
EVs

The first electric cars were developed in the late nineteenth century, with some of the earliest examples including Gustave Trouvé's electric car (1881), Flocken Elektrowagen (1888), and Ferdinand Porsche's La Toujours Contente EV that showcased 13.8 hp electric motors in each wheel (1898) in Germany. EVs became "popular" in the early twentieth century because they were simple, quiet, and clean, but their range was limited by the battery technology of the time. Steam-powered cars were also popular at the time and included the Stanley Steamer (1897), which set a land speed record in 1906, but they required frequent water refills in addition to refueling with fuels such as wood, coal, kerosene or gasoline. The first gasoline-fueled vehicles emerged with the invention of the ICE, and the first "modern" automobile was the Benz Patent-Motorwagen (1886) by Karl Benz. The gasoline ICE gradually became the dominant propulsion source, despite being noisy, unreliable, and polluting, because it could drive for longer ranges and refuel more quickly. Over time, advancements in the ICE and the availability of cheap gasoline led to the dominance of the ICEV throughout the twentieth century.

However, concerns about environmental impact and energy efficiency revived government interest in EVs in recent decades, and the recent evolution of EVs has benefited greatly from advancements in battery technology in addition to government incentives and regulations for cleaner air and higher efficiency, and funding for some charging infrastructure build-out. As technologies continue to evolve and costs decrease, EVs are becoming increasingly mainstream, offering a cleaner and more efficient alternative to the traditional ICEV. For many people, EVs are more fun to drive because they have faster off-the-line acceleration and quieter, smoother operation. There is also less maintenance because the brakes last longer, due to regenerative braking, and no oil changes are required. Tire tread might deteriorate sooner, however, because the vehicles tend to be heavier and can accelerate faster "off-the-line."

The evolution of battery technologies has been crucial to improving EV performance while lowering cost and offering longer range and durability. GM's EV1 two-seater, introduced in 1996, was one of the first modern EVs produced by a major automaker and had a range of up to 80 miles on a single charge (**Figure 3.1**). Early EVs, like the EV1, used lead-acid batteries because it was a well-established and affordable chemistry, widely used in starter batteries. However, it had several limitations, including low energy density (meaning there was only space to fit an 18.7 kWh battery, which supported a limited range of approximately 80 miles) and a short lifespan, which constrained the performance and practicality of EVs. Despite its innovative design and early interest, the EV1 program was discontinued due to limited demand and high production costs, but it demonstrated the viability and appeal of electric propulsion.

Due to the well-understood limitations of batteries, Toyota believed that an HEV made more sense for the consumer. The Prius, introduced in 1997, was one of the first mass-produced HEVs, combining a gasoline ICE with an electric motor and battery to improve fuel efficiency and reduce emissions. Like the EV1 and early Tesla models, it used a three-phase alternating current (AC) induction motor because it is relatively efficient and cost-effective. Although it is not a pure EV, the Toyota Prius was a groundbreaking vehicle that popularized the idea of electric propulsion and regenerative braking, and its success helped raise public awareness and consumer interest in electrified vehicles and accelerated the development of more advanced EVs (**Figure 3.2**). It used relatively expensive nickel metal-hydride (NiMH) batteries because they offered higher energy density, longer life, and better performance compared to lead-acid batteries.

GM's Autonomy concept vehicle was unveiled in 2002 and was the first to demonstrate the vision of a ground-up EV architecture that exploited the design freedom possible when replacing the ICE with electric propulsion, and the design freedom possible when mechanical and hydraulic braking and steering controls are replaced with electric by-wire controls (**Figure 3.3**). The result is a vehicle that has a completely flat chassis with no engine compartment and with electric motors at each wheel, enabling multiple bodies or coaches to be mated on top of the "skateboard," with a passenger compartment that has a flat floor and the freedom to move the seats, steering wheel, and pedals easily around the cabin. Eight months later, GM's Hy-wire was introduced as the first drivable vehicle to demonstrate the skateboard philosophy (**Figure 3.4**).

Figure 3.1 Cutaway of GM EV1, with 18.7 kWh battery in the tunnel, and the EV1 in motion.

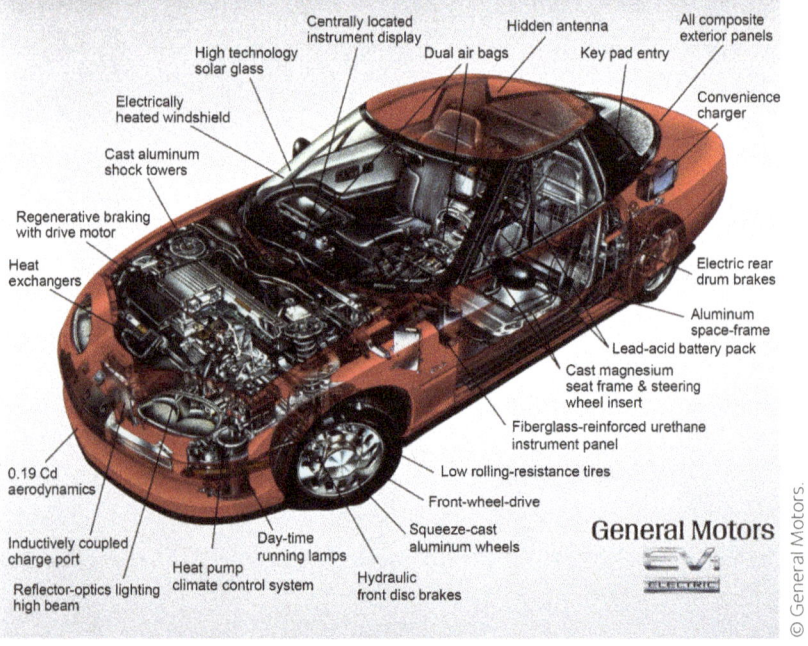

Figure 3.2 Toyota Prius, with 8.8 kWh battery pack behind the rear seat.

The original coachbuilders from the nineteenth century were artisans who built horse-drawn carriages and coaches by hand and often modified the body and interior to meet the specific needs and tastes of their clients. By envisioning a flexible vehicle platform that could have different body styles and interiors to suit different needs, such as a pickup truck, crossover, or commuter car, the Autonomy concept was a modern reinterpretation of this idea and created the potential for economies of scale because it would be possible to engineer and manufacture a common "skateboard" platform across multiple body types or even across multiple automakers (which will be discussed further in Chapter 8, Section 8.4.1). The concept showed the auto industry a vision for future electric mobility as well as for a new business model, which has now been realized as most automakers who build "ground-up" EVs (as opposed to those modifying one of their ICEV models) are using a skateboard approach or something similar.

Figure 3.3 GM Autonomy vision concept introduced the "skateboard" architecture (a battery pack can replace the hydrogen fuel cell system).

Figure 3.4 GM Hy-wire, the first drivable vehicle with a "skateboard" architecture.

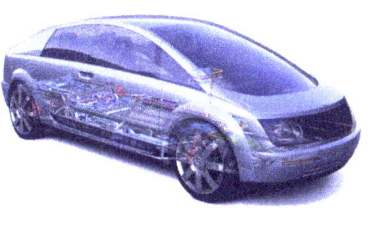

The Tesla Roadster was a pivotal moment in the history of EVs (**Figure 3.5**). Introduced in 2008, it was the first highway-capable EV to use lithium-ion battery cells and had a range of up to 245 miles on a single charge. Tesla engineered a 53 kWh battery pack by welding together 6831 of the cylindrical lithium-ion battery cells that were commonly used in laptop battery packs at the time (so-called 18650 because they are 18 mm in diameter and 65 mm tall) from 11 sheets or modules that each contained 621 cells. Although this approach was costly and complex to manufacture and surprised other automakers, it demonstrated Tesla's innovative approach and allowed them to enter the market sooner and with a high-performance and desirable vehicle, challenging the perception that EVs were sluggish and impractical.

Figure 3.5 Tesla Roadster, with the 53 kWh battery pack and power electronics occupying most of the trunk (and with no "frunk" to compensate!).

BYD (Build Your Dreams) was founded in 1995 as a battery manufacturer in Shenzhen (China) by Wang Chuanfu, a chemist, who initially focused on producing rechargeable batteries for mobile phones and other consumer electronics. In 2003, BYD entered the automotive industry by launching its first ICEV and then introduced its first EV, the e6, in 2009. In 2008, Warren Buffett's Berkshire Hathaway invested approximately $230 million in BYD, acquiring a 10% stake in the company, and BYD has also received significant support from the Chinese government [3.1]. It has since become the most vertically integrated automaker in the world and manufactures its own batteries, motors, electronics, and many other key components for all its EVs.

The Nissan Leaf was introduced shortly after the e6 and was one of the first mass-produced EVs that was designed from the ground up, rather than being a conversion of an existing ICEV model (**Figure 3.6**). The Leaf's successful launch and sales helped establish the viability of EVs in developed markets while the e6 performed a similar role in China.

Figure 3.6 Nissan Leaf chassis [3.2], with a 24 kWh battery pack under the floor.

At around the same time as the Nissan Leaf launched, GM introduced the Chevrolet Volt (**Figure 3.7**), the first EREV or extended range EV. As with a PHEV, it had a relatively small 16 kWh battery, sufficient for approximately 40 miles of driving. However, it also had a gasoline generator that could send electricity to the traction motor, recharge the battery, and be used to extend the range by another 300 miles or so. The EREV has the complexity of both an ICE and an electric propulsion system, which may make it a temporary or transitional solution until battery costs come down and a charging infrastructure becomes more ubiquitous. It is perhaps best suited for larger vehicles, like pickup trucks, that currently need very expensive and heavy batteries to provide sufficient range. Moreover, the extra electrical power that PHEVs and EREVs have onboard can make them attractive to contractors for powering electrical tools with their pickup trucks.

Unlike the early EVs, more recent EVs have tended to use permanent magnet (PM) motors because they are even more efficient and power-dense than AC induction machines. This is partly because the magnet provides a magnetic field without needing electric current, unlike induction motors that must induce a magnetic field in the rotor. Neodymium is often used as a permanent magnetic material because of its very high magnetic strength, but because more than 80% of the world's supply comes from China, there is an incentive for automakers to consider other materials, such as nickel, iron, cobalt, and ceramics, which have lower performance but a more secure supply chain.

The Tesla Model S, launched in 2012, was a tipping point for the EV industry because of its sleek design, long-range capability (more than 400 miles), and sportscar-like performance (**Figure 3.8**). Tesla's success with the Model S helped accelerate the adoption of EVs and inspired other automakers to invest in EV development, because it conclusively showed that EVs could be aspirational, practical, sporty, and luxurious, shattering many preconceived notions about EVs. It earned grudging respect from traditional automakers, although there were still doubts that Tesla could survive financially because it relied heavily on selling EV credits to other automakers to stay afloat.

Figure 3.7 2011 Chevrolet Volt chassis, with 16 kWh battery pack in the tunnel.

Figure 3.8 Tesla Model S chassis, with an 85 kWh battery pack mounted under the floor.

These concerns were laid to rest in 2018 when Tesla announced its first profit with help from the Tesla Model 3, which was introduced in 2017. The Model 3 was arguably the first mass-market, long-range, and relatively affordable EV, which has generated significant consumer interest and helped to accelerate the adoption of EVs globally.

The competition from Chinese automakers ramped up during 2023 with the introduction of the BYD Seagull, a compact electric hatchback, which immediately captured attention because of its low $11,400 price tag (**Figure 3.9**). The base version had a 30 kWh battery pack and provided 190 miles range, while the 38 kWh battery option increased the range to 251 miles on the Chinese Test Cycle (which has more frequent stops and starts than an equivalent test cycle in the US or EU, and more accurately represents typical Chinese urban driving conditions). While not luxurious, the interior boasts decent quality and offers a compelling package for budget-minded electric car buyers in China. European and US automakers are now striving to produce a sub-$25,000 EV, and BYD has recently overtaken Tesla as the world's largest EV maker.

Figure 3.9 BYD Seagull interior (base price $9700), interior and chassis with 39 kWh battery.

3.2. Microvehicles

In less affluent parts of the world and in more crowded cities, a smaller type of EV is gaining popularity, particularly in dense European cities where pollution, congestion, and parking are major transport issues. These so-called microvehicles can have two, three or four wheels, and the new generation are nearly always electric powered. They are smaller and significantly lighter than cars and tend to be driven short distances, which means that they only need small batteries and are already less expensive than equivalent ICEVs, especially when running "fuel" costs are included (**Figure 3.1**). Motorcycle taxi drivers in sub-Saharan Africa, for example, can spend around $2000 a year in fuel [3.3], and while a 1 kWh battery may only propel a typical car for two or three miles, it can power an electric scooter for up to 50 miles. Whereas gasoline can cost $0.03/mile for a motorcycle and around $0.10/mile for a car, the electricity cost for electric versions can be about ten times less [3.4]. This not only makes it much more affordable but also more profitable because nearly all the electric two-wheelers (E2W) are used for commercial purposes, with even Uber launching its first e-bike fleet in Nairobi in 2023 [3.5]. E2W and electric three-wheelers (E3W) are likely to be pursued aggressively in sub-Saharan Africa, the Indian subcontinent, and Southeast Asia as attractive replacements for conventional ICE motorcycles and tuk-tuks. One and a half million EVs, mostly E2W, were sold in India in 2023, roughly a hundred times as many as in Africa. However, led by Ethiopia and East Africa, it is likely that Africa will be the fastest-growing market for electric motorbikes. Unlike the passenger cars discussed in Chapter 2, Section 2.5, they will be new rather than used [3.5].

Table 3.1 EV types with typical specifications and current cost estimates.

	Example	Vehicle footprint (ft², approx.)	Vehicle mass lb, approx.)	Battery size (kWh, approx.)	Motor size (kW, approx.)	Range (miles)	Energy efficiency (Wh/mile)	Vehicle price $ (USD, approx.)
Electric bike	Currie Ezip Trailz	10	30	0.25	0.5	15	16	700
Electric tuk tuk	Mahindra Treo SFT	30	750	7	8	80	92	4000
Electric quadricycle	Citroen Ami	28	1100	5	6	45	110	8000
Electric car	Tesla Model Y	85	4500	70	300	315	247	43000
Electric pickup truck	Ford Lightning Platinum	125	7000	110	350	255	513	93000

© SAE International.

Microvehicles also include the license-free quadricycles found in Europe and the neighborhood EVs in the US. They look like cars but are not as crashworthy and, with a top speed of approximately 45 mph, they are not designed for operating on highways. Despite these limitations, their costs are often still relatively high because they are mostly sold in developed countries to affluent consumers and are produced in low volumes by small firms and startups with sub-optimal supply chains. However, as automakers such as Stellantis and Renault begin building them at volume, their cost could be well under $10,000, which is significantly less than a car.

To facilitate the adoption of microvehicles optimized for specific use cases rather than serving as multipurpose vehicles, a fundamental shift in consumer mindset and expectations may be required, particularly in the developed world. Historically, personal vehicle ownership has been driven by the perceived need to own a single vehicle that can handle virtually every conceivable driving scenario, from daily commuting to road trips to hauling and towing tasks. As covered in Chapter 2, Section 2.1 this means they are vastly overengineered and more costly than necessary. However, consumers may need or want to embrace a more pragmatic approach of using purpose-built EVs tailored to their predominant mobility needs, such as daily commuting to work and running errands around town, while using public transport or subscribing to a vehicle service for longer journeys, as discussed in Chapter 7. The design of tailored, or right-sized, micro-EVs is the subject of Chapter 8, Section 8.4.

3.3.
EV Technology

Steady progress has been made over the last decade to improve EV cost, performance, and utility. This has largely been achieved by reducing the cost, size, and weight for each unit of energy stored in the battery. A battery pack contains not only battery cells but also a battery management system (BMS), a cooling system, wiring and connectors, and a protective case. As a rough rule of thumb, a complete battery pack might cost 1.3 times as much as the total cost of the battery cells inside it. In 2024 the average battery pack cost was $115/kWh [3.6] and is expected to continue declining because of economies of scale, technical progress, and improved manufacturing processes to below $100/kWh, a key threshold for widespread EV adoption because it means EVs should have a similar purchase price as gasoline ICEVs. In fact, BYD has already driven the more affordable LFP cell prices down to below $60/kWh and is targeting $44/kWh with increased scale and vertical integration [3.7]. Battery materials are approximately 70% of this total cost, equipment and processing make up nearly 20%, and the remainder is labor cost. The progress in energy density, charging rate, and battery life has come about through a variety of improvements, such as electrode structures that can store more energy and higher voltage, more durable and nonflammable electrolytes, and advanced BMSs that allow faster charging without battery degradation. **Figure 3.10** (left) shows an energy storage comparison, by weight and volume, for different battery chemistries while **Figure 3.10** (right) illustrates the approximate ten-fold loss of energy performance that typically occurs between theory and actual mass-produced battery packs which means that claimed laboratory breakthroughs should always be treated with caution. There is also typically a decade gap between a promising battery chemistry being discovered in the lab, at the Wh scale, and mass production of the same battery in the plant, at the GWh scale.

Figure 3.10 (Left) energy content of different battery chemistries, and (right) typical losses that occur when scaling from theory to the vehicle battery pack [3.8].

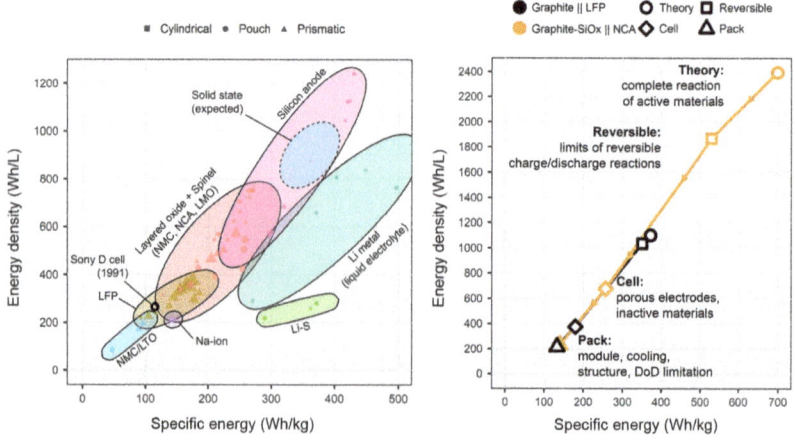

Frith, J.T., Lacey, M.J. & Ulissi, U. A non-academic perspective on the future of lithium-based batteries. Nat Commun 14, 420 (2023). https://doi.org/10.1038/s41467-023-35933-2.

Energy content is not the only metric to compare battery chemistries, as shown in **Figure 3.11**, because safety, cost, and durability are also very important. Batteries can become cheaper because of economies of scale and learning curve improvements with existing batteries (such as with LFP, and lithium nickel manganese cobalt, [NMC]), or because of a new, innovative battery chemistry that offers higher performance and uses less raw material (such as lithium-sulfur or solid-state lithium batteries). Future batteries could also become cheaper because they use more widely available raw materials, albeit with some performance loss. Sodium-ion batteries not only reduce dependence on lithium (sodium is more than 1000 times as abundant as lithium) but also on nickel and cobalt, and because it can use the same production processes as are used for lithium-ion batteries they could be made in the same plant. CATL, the world's largest battery manufacturer, has recently produced a sodium-ion battery for evaluation. These batteries are less energy dense, but this may not be a barrier to usage, particularly for urban EV fleets. However, because the price of lithium has dropped over 80% between 2022 and 2025, the economic attractiveness of sodium-ion has become more challenging, but this could change in the future.

Figure 3.11 Various lithium-based battery chemistries, with associated attributes (performance improves from the center outward) [3.9].

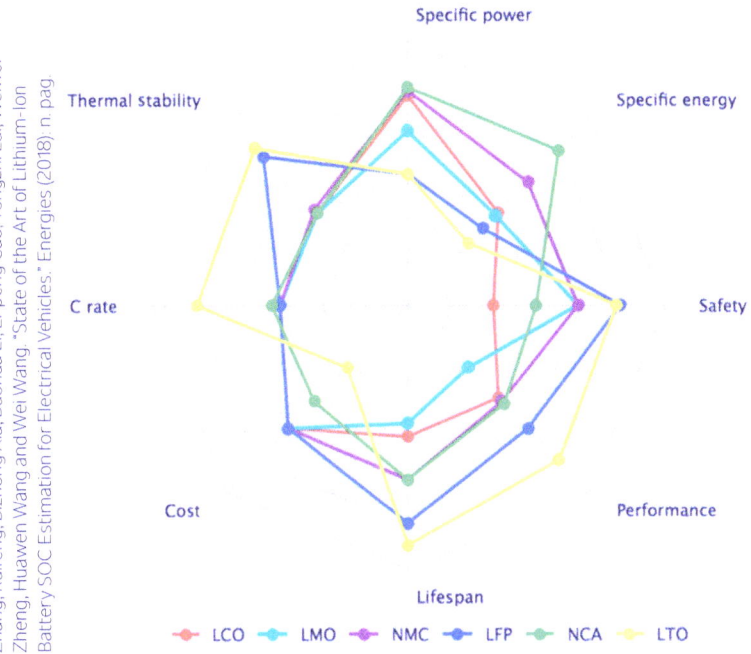

Future battery chemistry energy density improvements could translate into vehicles having twice as much range as current EVs, which would make them competitive with most ICEVs. In 2023, for example, a Chinese EV maker, Nio, drove its ET7 vehicle 650 miles (1044 km) in a live-streamed event, and with a measured range of 621 miles (1000 km) on the China Light-Duty Vehicle Test Cycle [3.10]. Samsung SDI, a battery supplier to Hyundai, Stellantis, and GM, has also announced plans to commercialize a solid-state battery that will provide EVs with 600 miles of range in 2027 [3.11].

In addition to continuous battery technology improvements, another area of significant innovation in the electric mobility space is EV charging. Typical AC charging from a power outlet needs to be converted to direct current (DC) by the vehicle's onboard charger before the battery can be charged whereas DC charging delivers DC directly to the battery and gets around the limited power capacity of the onboard charger. DC fast chargers also communicate directly with the vehicle's BMS to optimize the charging rate and protect the battery. However, DC fast charging systems must use thicker cables, more robust connectors, and advanced cooling systems to handle the higher current and heat produced from power losses.

DC fast charging (level 3 charging) at rates above 350 kW, enabled by higher voltage electrical systems inside the vehicle, allows for rapid EV charging, and CATL has shown it can add 370 miles of range to an LFP battery after charging for ten minutes, equivalent to adding more than 2000 miles of range per hour. Not to be outdone, a few months later BYD announced it could add up to 250 miles of range in five minutes with 1000 kW charging, comparable to refueling a gasoline ICEV. This is more than twice as fast as a Tesla Supercharger, which can only add up to 15 miles of range to a Model 3 each minute. BYD has also announced plans to roll out 4000 fast charging stations across China [3.12]. It can be argued that fast charging at this rate can be as effective in reducing range anxiety as increased vehicle range.

Most automakers favor continued improvements in the charging infrastructure to enable long-distance travel and to address range anxiety concerns for EV owners, but in China, at least, there is some competition from battery swapping. Although battery swapping requires standardized battery packs and the swapping stations are costly to set up and maintain, Nio has established a network of more than 2651 battery swapping stations in China and another 58 in Europe, primarily in major cities, that can swap batteries in under three minutes (**Figure 3.12**) [3.13]. Nio has stated that a swap station needs to service a minimum of 60 to 70 vehicles a day to break even [3.13]. The battery swap model can make EVs more affordable to purchase because the cost of the battery can be eliminated from the initial vehicle purchase price, and it could make EVs more efficient to run because a smaller, lighter battery could be used most of the time with larger batteries being swapped in only as needed. It can also reduce strain on the electric grid and extend battery life. Newer, more advanced batteries could even be swapped in as a hardware "update" over time, decoupling the vehicle from the battery. CATL is beginning to roll out "choco-swap" standardized batteries that are designed to be easy to swap out, with a goal of installing 1000 swapping stations in China in 2025 and eventually 30,000 across the country [3.14]. Perhaps China will be an exception.

Figure 3.12 One of Nio's battery swap stations (Chongqing, China).

Unlike ICEVs, which tend to evolve slowly over time and may hardly evolve at all in the future because of limited research investment, progress in EVs seems to occur on an almost weekly basis. There are also encouraging developments that promise to further improve range, cost, recharging time, and recyclability, which are all needed to address affordability and sustainability. These will be covered in Chapter 4, Sections 4.1 and 4.2.

3.4. Connectivity and Software-Defined Vehicles (SDVs)

The history of connected vehicles is much shorter than for EVs, but recent developments also trace back to the 1990s (**Figure 3.13**). Connected vehicles is a broad term and includes vehicle connectivity with the smartphone (such as Bluetooth pairing, ride-hailing, locating EV charging stations, and remote start), as well as the vehicles having connectivity themselves for features like GM's Onstar, satellite radio, remote diagnostics, and infotainment.

Figure 3.13 Evolution of connected vehicle features and services.

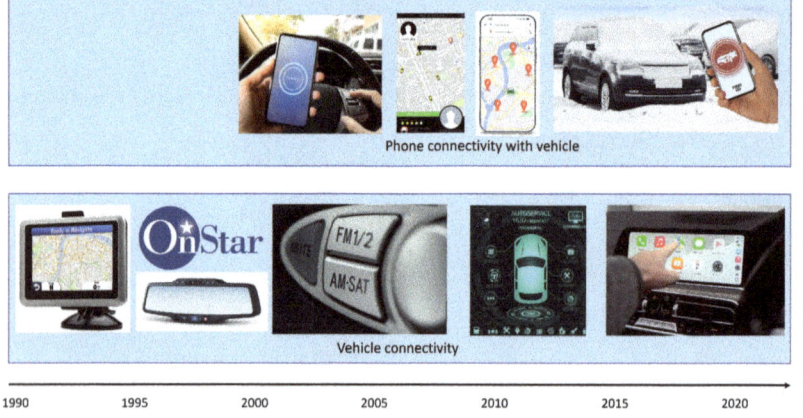

Connected vehicles began with a focus on safety and then evolved to encompass vehicle diagnostics before expanding into consumer infotainment and convenience features. In the future, they will enable payment capabilities ("wallet on wheels") and will be key to supporting road safety and traffic flow with vehicle-to-everything communications (V2X), to be discussed in Chapter 4, Section 4.7. As with ride-hailing, connectivity is essential to mobility-as-a-service (MaaS), the topic of Section 3.7.

In the mid-1990s, GM introduced OnStar on its vehicles. The primary focus was safety and security with features like automatic crash response and emergency assistance. At a time when most people did not have a cell phone, GM's Onstar was able to make cellular calls even in remote locations where a cell phone would not work (because it had a larger, more powerful antenna that was integrated into the vehicle). Even if the driver was incapacitated during a collision, the vehicle would know based on the airbag deployment and GPS location, and would be able to call local emergency responders. Connected vehicles can transmit data not only about accidents and breakdowns but also about potential security threats, such as theft—allowing automakers to respond quickly, even shut down stolen vehicles, and improve overall safety and security. Several other automakers, including Toyota and BMW, followed shortly after GM and with similar offerings.

As connectivity became more prevalent with cell phone adoption in the early 2000s, automakers recognized the potential for connected vehicles to provide enterprise benefits, particularly in the areas of vehicle diagnostics, maintenance, and fleet management. Connected vehicles can transmit real-time data about the vehicle's performance, component health, and diagnostics, and this allows automakers to collect real-time vehicle diagnostic data which can be shared with dealerships so that the vehicle arrives at the repair shop at the same time as the replacement part and as the technicians become available. For commercial fleet owners, connected vehicle data not only allow predictive maintenance scheduling but also efficient fleet management, such as real-time vehicle tracking, driver monitoring (and training), and route optimization.

Around 2010, when smartphones took off, automakers began leveraging connectivity to offer infotainment systems (combining entertainment, navigation and vehicle information), as well as remote control and monitoring features, such as lock/unlock, remote start and cabin heating, ventilation, and air conditioning (HVAC), and vehicle location tracking. Automakers can now collect and analyze data on vehicle performance and on how customers actually use their vehicles, including their driving patterns and what features they use. These telematics data can be used not only for predictive maintenance and for service scheduling but also to improve future vehicle designs by informing automakers of feature usage and component durability, for example. Connected vehicles also allow automakers to deliver software updates and firmware upgrades OTA, without the need for the car or customer to visit a service center, enabling convenient updates to infotainment systems, ADAS, and even core vehicle functions. Connected vehicle data also allow automakers to personalize the in-vehicle experience based on user preferences, location, and driving behavior and provide personalized infotainment settings, customized navigation routes, and tailored recommendations.

The need to improve cybersecurity may lead some automakers, particularly luxury brands, to consider embedding two cellular modems in some vehicles to provide greater separation for safety critical systems. The main cellular connection, common in almost every new car, is used for critical services such as OTA updates and emergency services, as well as for collecting data to send back to the automaker while a second modem might be used for streaming infotainment-related content and to provide redundancy.

Connected vehicle data may also generate new revenue streams for automakers, such as subscription-based services, usage-based insurance models, and data monetization opportunities. This combination of changes, driven by increased data collection and connectivity to the cloud, has recently given rise to the umbrella term "software-defined vehicles," a term created by engineers when "software vehicles" (SVs) might have been more consistent with EVs and AVs (**Table 3.2**).

Table 3.2 Traditional automotive and SDV business models.

Feature	Old business model	SDV business model
Focus	Selling physical vehicles	Software-enabled hardware platform
Revenue streams	One-time car sales, after-sales services	Care sales, software subscriptions, in-app purchases, data services
Product cycle	Slow, hardware-driven innovation cycles	Faster, software-driven innovation cycles
Customer relationship	Transactional	Ongoing, focused on engagement
Challenges	Limited recurring revenue, supply chain dependence	Culture shift, tech expertise, potential manufacturing changes

SDVs will bring new features and business models to customers, as well as change how automakers are organized and how they develop new vehicles. As discussed in Chapter 2, Section 2.2, the automotive industry has already shifted from a focus on mechanical hardware toward electronics, and now it is taking this one step further and reorienting toward emphasizing software, which allows for the development of new features and a greater degree of automation. This has significant implications for the automotive supply chain because many automakers are looking to develop the software themselves and leaving the suppliers to simply provide "dumb" hardware, instead of bundling software with an individual ECU as they have previously done. By developing the software in-house, automakers hope to copy Tesla and monetize updates without relying on supplier coordination for software support. After all, even a "simple" change to a semi-autonomous feature, for example, will involve modifications to the propulsion ("lift the throttle"), chassis ("apply the brakes"), and infotainment ("notify the driver") systems that have traditionally been developed by separate suppliers.

SDVs can collect gigabytes of data every hour [3.15] because they contain more data-rich sensors, such as cameras, and have greater connectivity to the cloud, via Wi-Fi or cellular communications. These data can be leveraged to create new features, such as personalized infotainment or vehicle settings, and to provide more security by supporting biometrically authenticated payments. The cloud plays a crucial role in supporting SDVs because it enables the collection, storage, and analysis of large amounts of data generated by the vehicles for the purpose of understanding vehicle performance and customer behavior, giving automakers a chance to quickly iterate and update software with OTA features and improvements. The cloud can also support vehicle development by using real-time data from vehicle components and seeing the effect of an OTA update on the new performance. This real-time data comparison can even be used to improve simulation accuracy when designing new solutions. The cloud has also allowed automakers to scale their infrastructure based on demand and to pay for additional resources only as needed, reducing upfront costs and improving operational efficiency. However, a significant challenge with using the cloud is data management because unlike traditional vehicle sensors that sense tire pressure, wheel speed, engine coolant temperature, and so on, modern cars with ADAS sensors (especially cameras and lidar, or light detection and ranging) can generate massive amounts of data and the cost to upload the data for storing and analysis in the cloud is expensive. For this reason, uploaded data should be edited to reduce the file size and bandwidth requirements. Another challenge is that wireless networks can be slow or even nonexistent in some areas and so smarter, more concise ways to send an OTA software update are being developed.

Last, but certainly not least, real-time connectivity is also enabling the transition from vehicle ownership to usership, which will be covered extensively in Chapter 7.

3.5. AVs

The vision of AVs has fired people's imagination for decades but, until recently, seemed so far off as to be science fiction because driving a vehicle requires us to perceive the environment, plan the path to take, and execute the vehicle controls for braking, steering and accelerating. This is difficult, to say the least, and requires that an AV needs to do the following:

- Collect data from various sensors to perceive its surroundings, and using advanced software, classify the objects, such as pedestrians, vehicles, traffic signs, and lane markings. The software must then determine the vehicle's precise location and orientation within the environment.
- Based on the perceived environment, the software must plan a feasible path from the vehicle's current location to the desired destination. Appropriate maneuvers, such as a lane change, merge, or yield, that are based on traffic rules, and the predicted movements of other objects and their intended path must then be generated before the software decides on a smooth and dynamically feasible trajectory that considers vehicle constraints and any obstacles in the path.
- This planned trajectory must then be converted into low-level control commands for the vehicle's steering, throttle, and brakes to execute the desired motion. The software must continuously monitor the vehicle's actual motion and surroundings, and continually be ready to adapt and make necessary adjustments.

This process is continuous and cyclical, with the "vehicle," or more accurately the vehicle's autonomy system, constantly perceiving its environment, updating its plans, and executing the necessary control actions.

The first major effort to develop AVs, or more accurately a Level 2 AV (see Chapter 4, **Figure 4.15**), was Project Prometheus, a pan-European research project that was launched in 1986 (**Figure 3.14**), and was spearheaded by Ernst Dickmanns. It was completed in 1995 when an automated Mercedes S-Class drove from Munich to Copenhagen and back, a round-trip journey of more than 1000 miles. Because GPS was not available at the time the vehicle needed to rely principally on four different cameras to monitor the environment and up to six other vehicles, but it still hit speeds above 100 mph and drove autonomously in real-world highway traffic, albeit with a safety driver. Image processing power was also very limited at the time and so some shortcuts were necessary, such as focusing on object edges rather than the whole object, and on changes from frame-to-frame, and these shortcuts are still being used today to compress data and reduce processing time. The development vehicle for the project, a Mercedes-Benz Vario van, was chosen because it had ample interior space to house the computer and electronics systems, even though one smartphone today has more computing power! Project Prometheus has been credited for accelerating the development of several ADAS features, such as AEB and LDW, as well as for stimulating the development of the DARPA Grand Challenge and Urban Challenge events in the US a few years later [3.16].

Figure 3.14 A 1995 autonomous Mercedes-Benz S-Class (Project Prometheus) with the autonomy system components (no radar or lidar!) [3.16].

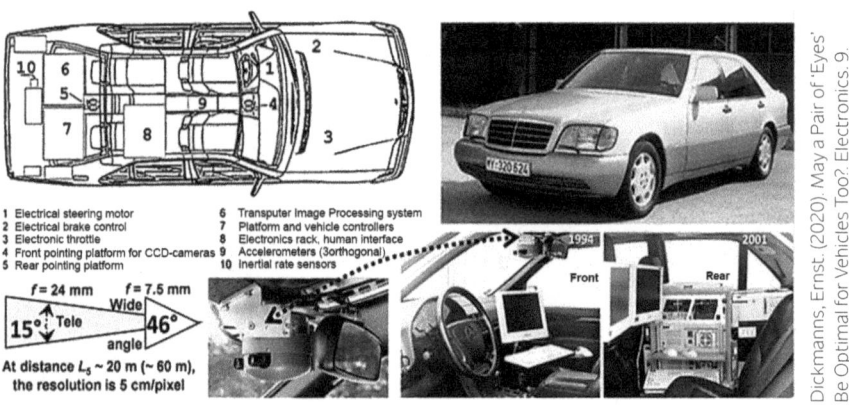

At around the same time, during the mid-1990s, GM explored various guidance systems for AVs, including magnetic markers embedded in roadways. Its vehicles were equipped with radar to maintain a safe following distance and to demonstrate the increased road throughput possible with closer vehicle platooning. In 1997, GM tested a magnetically guided vehicle on a 7.6-mile (12.2 km) stretch of Interstate 15 near San Diego, California. This automated highway system used nearly 93,000 magnetic markers buried in the road to provide guidance and positioning information to the vehicles [3.17]. The magnetic markers were installed in the center of the lane, and the vehicles were equipped with sensors to detect and follow the magnetic signals. This allowed the vehicles to maintain their lane position and navigate autonomously, even in poor weather conditions or when lane markings were obscured. While the magnetically guided system demonstrated promising results in controlled environments, it faced several challenges for widespread implementation, such as the high cost of installing and maintaining the magnetic markers on a large scale, the inflexibility of the system to handle dynamic changes in the environment, and the potential interference from other sources of magnetic fields.

Fortunately, sensing technology improved to the point where needing to rely on the road infrastructure for autonomous driving became a nonissue. Although radar was invented in the early twentieth century, it was not until 1999 that the world's first radar-based ACC system was commercially introduced, by Mercedes-Benz. This system could detect the distance and speed of the car ahead, and automatically adjust the vehicle's speed to maintain a safe following distance. In 2003, Volvo became the first automaker to use radar for detecting blind spots, and cameras started to be introduced in automobiles a few years later because they offered a couple of significant advantages over radar, namely that they can distinguish objects (such as cyclists, motorcyclists, and cars) and they can provide a visual display. In 2011, Mobileye supplied the first forward collision warning system to several automakers, and this relied on a single camera mounted inside the windshield.

With continuous performance improvements in radar and camera technologies, as well as proven cost reductions that have been driven by mass economies of scale, it became possible to imagine AVs becoming viable without a need for costly road infrastructure changes. A milestone in AV development was achieved on November 3, 2007, when the DARPA Urban Challenge took place at an unused Air Force Base in Victorville, California. This setting mimicked a real suburban/urban environment with traffic signals, intersections, and even pedestrians (played by actors), in stark contrast to previous DARPA Grand Challenges that took place in the desert. AVs had to complete a 60 mile urban environment course while obeying traffic laws, merging into traffic, avoiding obstacles, and interacting safely with other vehicles on the road without human intervention. Carnegie Mellon University's self-driving vehicle, "Boss" Tahoe, supported by GM, Continental, and Caterpillar, won the event by having no collisions and finishing the course in the fastest time (**Figure 3.15**). The DARPA Urban Challenge is often credited with jump-starting industry focus on AV development, particularly as it stimulated Google to start their program "bet" in 2009, which became closely watched by the auto industry.

Figure 3.15 Chevrolet "Boss" Tahoe (winner of 2007 DARPA Urban Challenge) had 11 lidar sensors, five radar sensors and two cameras, plus a trunk full of hardware for data processing to support autonomous operation.

The DARPA Urban Challenge demonstrated for the first time that it was feasible for AVs to operate safely in a complex urban environment, and it also set the direction for the hardware approach that has become standard for most AVs ever since. The leading teams all used a similar sensor suite for perceiving the environment and any obstacles. This relied on combining several cameras and radars with lidar, and this approach has become the *de facto* standard for most true AVs. The argument given is that each sensor has complementary strengths and limitations:

- Cameras provide high-resolution visual data, essential for tasks like lane detection, traffic sign recognition, and object classification. In a sense, they mimic eyesight and can capture detailed information about the environment.
- Radar is useful for detecting and tracking moving objects, such as vehicles and pedestrians, even in challenging weather conditions

like rain or fog. They can also provide accurate distance and velocity information, which is necessary for safe navigation.
- Lidar can create a detailed 3D map of the environment by using laser beams to measure the distance to surrounding objects. These 3D data are essential for tasks like obstacle detection, path planning, and localization (determining the vehicle's position relative to its surroundings), in essence combining some radar and camera capabilities, and providing some redundancy.

The combination of these three sensor types provides a comprehensive and complementary view of the environment, enabling robust perception and decision-making capabilities for AVs (**Figure 3.16**) and even if the quantity and placement of these sensors vary from one AV to another, the core sensor suite of cameras, radars, and lidar has become accepted as both necessary and sufficient, even though a viable AV has yet to reach the level required to satisfy developers, customers, and government regulators. There are some dissenters to this generally accepted sensor suite. Tesla, for example, believes that cameras alone may be sufficient whereas several companies believe the three sensor modalities are insufficient, and that it is necessary to add infrared sensors, particularly to improve operation in poor lighting conditions, such as at night. As mentioned in Chapter 4, Section 4.7 some AV developers in China are also looking to integrate V2X as an additional "sensor."

Figure 3.16 Typical sensors and sensor locations for AV operation.

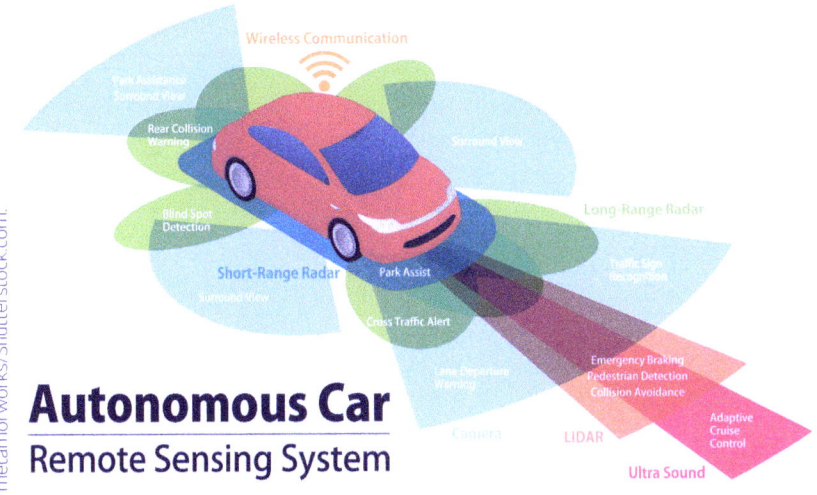

The next breakthrough in AV development occurred with machine learning progress in 2012. A research team led by Geoffrey Hinton at the University of Toronto, working with Google, Microsoft, and IBM, used a deep learning algorithm to significantly reduce the error rate in the ImageNet image recognition challenge [3.18]. This success demonstrated the potential of deep learning to solve complex tasks like image recognition. Convolutional neural networks (CNNs) have achieved impressive results in image recognition, enabling accurate detection and classification of objects, pedestrians, and traffic signs, and they are now widely used in AVs for perceiving the environment via input from the cameras. Other methods, such as deep reinforcement learning and generative adversarial networks, have also helped to improve the training of the AVs to better navigate through complex scenarios, predict the behavior of other road users, and recognize the environment.

In addition to better sensors (increased range, resolution and durability, lower cost, etc.) and more sophisticated software, there have also been significant gains made in the compute platform that is needed to process the vast amounts of incoming sensor data and to instantaneously predict what other road users are about to do, what path for the AV (or ego vehicle) to plan, and what actions the AV should take in terms of braking, steering, and accelerating. These improvements include faster processing through a greater number of cores, a mix of different types of semiconductors for optimizing each software step, greater redundancy and fault tolerance, and 5G connectivity for real-time data sharing with the cloud, all while minimizing power consumption and heat rejection.

In contrast to electric and connected vehicles, however, we still do not have AVs that are commercially available to buy or use. This may change in the near future, given the technical progress that is being made and by applying such progress to easier, more constrained use cases, such as lower speed zones (e.g., shuttles or taxis on campuses, or city centers) and on "quasi"-dedicated highway lanes with minimal lane changes (e.g., goods delivery trucks). In the meantime, autonomy continues to stimulate the development and commercialization of simpler ADAS that aim to reduce passenger vehicle crash frequency and severity by detecting collisions and applying appropriate evasive action sooner than a human driver. Emerging developments with ADAS will be discussed in Chapter 4, Section 4.8.

3.6. Autonomous, Connected, Electric Mobility

One reason for the intense development of autonomous, connected, and electric vehicles is that they support or reinforce each other when combined, which may not be as evident with ICEVs. For example, AVs need significant amounts of power for processing the sensed data, and this is easier to deliver with EVs; they also benefit from connectivity for remote operations, OTA updates, and, possibly in the future, vehicle-to-everything (V2X) communications (see Chapter 4, Section 4.7). EVs also benefit from connectivity to find EV charging stations, for example, while autonomous operation could make automated or wireless EV charging even more convenient. In short, the three technical trends work together and can also make MaaS and integration with public transport easier to achieve.

Three GM EN-V (Electric Networked-Vehicle) concept vehicles were unveiled for the *2010 Shanghai World Expo,* whose theme was "Better City, Better Life" (**Figure 3.17**). Such a vision could lead to a city center with fewer parking lots, better air quality, quieter streets, no traffic accidents, and faster speed of movement across the city for improved economic productivity while using less energy. The EN-V was significant because it was the first vehicle to combine and demonstrate all three trends—electrification, connectivity, and autonomy—in a social or side-by-side, two-passenger form that was designed to address the major challenges facing urban mobility (reducing energy usage, air pollution, road accidents, traffic congestion, and parking space). The EN-V featured autonomous driving capabilities, vehicle-to-vehicle (V2V) communications, and a battery electric powertrain with electric motors in each wheel, showcasing GM's vision for future transportation. It was driven extensively by the media in 2010 and demonstrated videoconferencing between passengers in two EN-V pods that were platooning together (with a vision that the pods could become a "virtual" large vehicle when platooning and then be able to separate into individual pods as needed). Taken even further, this platooning concept suggested the potential for improving the energy efficiency and flexibility of public transport, particularly outside the city center and at night when trains and buses are mostly empty.

Figure 3.17 GM EN-V concept vehicles were the first driveable vehicles to demonstrate the "now-accepted" convergence of autonomy, connectivity, and electrification. They could operate both outdoors and indoors due to their zero emissions and highly maneuverable operation. The vehicles contained two antennaes (for GPS and for V2V communications), a forward-facing camera, and an electric platform with independent wheel motors.

Other futuristic concept vehicles were demonstrated shortly afterward, such as Nissan's PIVO 3, which had an electric powertrain, autonomous driving capabilities with a 360° spherical cabin design, and connectivity features for communication with infrastructure and other vehicles. Mercedes-Benz's F 015 Luxury in Motion concept was unveiled at the Consumer Electronics Show in 2015 and had connectivity features for infotainment and an advanced HMI with gesture control and augmented reality displays. While these and other concepts and prototypes showcased the integration of autonomy, connectivity, and electrification, it should be noted that they were not mass-produced vehicles available for purchase by consumers but were designed to demonstrate the potential of these technologies and their convergence in future mobility solutions.

The first widely available production vehicle that combined all three elements of autonomy (limited self-driving capability), connectivity (Internet access and OTA updates), and electrification (battery electric propulsion system) was arguably the Tesla Model S with the introduction of "Enhanced Autopilot" around 2016. The seamless integration of advanced autonomous driving, comprehensive connectivity, and battery electric propulsion in a mass-produced vehicle is an ongoing process, however, as automakers, automotive suppliers, and technology companies continuously advance these technologies and search for attractive new business models.

3.7.
MaaS

A common acronym for future mobility used in the automotive industry is ACES (or CASE). The "S" stands for shared mobility. Shared mobility has actually been around for a long time, especially when one considers the horse-drawn carriage, for example. However, it was not cheap because a horse-drawn carriage journey in the nineteenth century would probably have cost a few hours of labor for each mile of the journey, so that a trip covering 100 miles may have cost a month's wages while taking one to two days.

The first public bus system was introduced in London in 1829, the first intercity train system connected Liverpool to Manchester in 1830, and the London Underground became the world's first subway in 1863. The first gasoline-fueled taxicab, known as the Daimler Victoria, was built by Gottlieb Daimler in 1897 and began operating that year in Stuttgart. Shortly afterward, gasoline taxis were introduced in other major

European cities and then in New York City in 1907. These early taxis were an obvious evolution from horse-drawn carriages.

In the early twentieth century, rental car services, such as Avis and Hertz, emerged to allow customers to rent vehicles for shorter periods of time, typically for a few days or weeks. Over time, the service has allowed more flexibility in rental car pickup and drop-off locations, with airports being the dominant location, and more choice of vehicle types, but the costs can be high, the paperwork can be time-consuming, and there can be limited availability during peak travel periods. To address some of these issues, carsharing services, such as Zipcar and Car2go, began in the late 1990s to provide access to shared vehicles for shorter periods of time, usually by the hour or minute. Carsharing services can be station-based, where users pick up and return the car at designated stations, or free-floating, where users can pick up and drop off the car at different locations. Carsharing became somewhat popular in European cities, but the business has recently been hurt by the rise of ride-hailing, exemplified by Uber and Lyft, which has made "taxi" rides more affordable for many and reduced the appeal of carsharing as an alternative to renting a car, while also allowing mobility for people who cannot drive.

Ride-hailing has had an even more significant impact on the taxi business. The disruption to the taxi industry is obvious from the loss in value of a taxicab medallion, which is a transferable license that is required to operate a taxi in certain cities. Before ride-hailing existed, medallion systems artificially limited the supply of taxis to avoid overcrowding the market, and this meant they were valuable and could cost over $1 million in cities like New York and Chicago. However, Uber and Lyft bypass the medallion requirements by using personal vehicles that are driven by nonprofessional drivers, and the effect has been to flood cities with ride-hailing "taxis" to the extent that the average number of daily taxi rides in New York City halved from nearly 500,000 in 2012 to around 230,000 in 2019 [3.19]. Not surprisingly, the value of medallions has dropped by around 90% in New York City since the introduction of Uber and Lyft [3.20].

Another side effect of connectivity and smartphones is that real-time navigation is "free," and this has reduced the need for a taxi driver to have extensive knowledge of the area and to know shortcuts to save time when road conditions change. Historically, one of the key requirements to obtain a London taxi driver's license was to pass an extremely rigorous test called "The Knowledge," which involved spending several years traveling around London and memorizing the layout of London's archaic and complex street system. With the likes of Google Maps becoming ubiquitous, any ride-hailing driver can now immediately navigate the quickest way from point A to B, leveraging the crowd-sourced real-time data from other road users. In many respects, this is even better than "The Knowledge" because it is more up-to-date with construction or road accident information. Paralleling taxi developments elsewhere, there were 1016 London black-cab taxi drivers in 2016, but only 104 licenses were issued in 2024. This 90% decline has led to predictions that they will disappear completely by 2045 without government intervention [3.21].

Some ride-hailing services, such as SIXT, have also integrated taxi drivers into their network, and the app connects passengers with taxis and licensed cab drivers who are part of the app's network. This provides more customer choice and allows passengers to book and pay for taxi rides through the app instead of hailing a cab off the street. The app also tracks the taxi's location and route, provides price information before taking the ride, and allows payment with a credit card, innovations that were first introduced by Uber. The benefit of taking a taxi, even though it may be more expensive, is that the drivers and vehicles tend to be more highly regulated with stricter driver training and vehicle maintenance requirements, and some taxis may also have disability access.

Ride-hailing services, such as Uber, Lyft, and DiDi, have allowed users to conveniently book and pay for on-demand transportation services through a mobile app or online platform so that riders can order a vehicle to pick them up at a specific location, and the driver will drive them to their destination. The service benefits from wide availability in many cities around the world with convenient, cashless transactions and real-time tracking, but surge pricing makes it unreliable during peak hours or high-demand periods, and coverage can be nonexistent in suburbs or rural areas late at night or early in the morning.

Peer-to-peer sharing is also currently available (e.g., Turo, Getaround), but it has not yet taken off to the same extent as Uber or Lyft. The typical concerns that vehicle owners have with sharing their vehicle is that it may not be returned on time, may be damaged, or may not be clean. Vehicle users, on the other hand, face the inconvenience of having to pick up and drop off the vehicle, perhaps having to find transport in the process. Some of these challenges, perhaps all, may be solvable. Autonomy, for example, can solve the pickup and dropoff challenges, but until the challenges of insurance and cleaning are solved for the owner, peer-to-peer sharing may remain a niche business. If, at some point in the future, the owner has an AV then it may be possible for it to drive directly to the customer pickup point, avoiding collisions in the process, and even drive to a cleaning and recharging service before returning to the owner. This will be discussed more extensively in Chapter 7, Section 7.4. This type of autonomy is easier to achieve than in a robotaxi because there can be more flexibility to avoid complex routes, there are no occupants, and the user could even choose to pay more for faster vehicle delivery service. Car owners will not only generate revenue but will also save on parking fees. Alternatively, a fleet could manage all aspects of the business and share the revenue with the car owner in return for not having to buy and own the vehicle.

Although it does not go "door-to-door," public transport is often the most effective solution in urban areas. Addressing the "last mile" is where MaaS comes in. MaaS is a broader concept than public transport integration, or multimodal integration, because it encompasses both public and private transportation service options. Public transport integration has traditionally focused on combining personal vehicles (like cars or bicycles) with various types of public transport, and the focus has been on physical infrastructure improvements such as park-and-ride facilities or bike storage at train stations or on the front of the bus. The focus has not been on creating a unified digital platform or payment system.

Both MaaS and public transport integration share an aim to integrate different transportation modes and to reduce private car dependency. However, public transport integration is limited to public transport and is usually publicly managed, whereas MaaS includes both public and private options and has greater private sector involvement and different business models involving more partnerships and payment models. It offers more personalized, flexible solutions compared to the fixed routes and schedules of integrated public transport because it typically includes the following solutions:

1. Public transport integration: A unified system that allows seamless transfers and payments between bus, train, subway, and other public transport options.
2. Mobility services: Access to carsharing, vehicle subscriptions, and ride-hailing services that can provide door-to-door alternatives to private car ownership.
3. Micromobility: Access to shared e-scooters, bikes, and e-bikes while also supporting comprehensive bike lane networks and bike-friendly policies.

Historically, there have been few alternatives to driving other than cycling, taking a taxi, or using public transport, but mobility services and micromobility have recently offered more choices. Shared vehicle services include cars, bikes, and e-scooters, and they can provide access without ownership, reducing the need for personal car usage. However, carsharing has recently declined in popularity as ride-hailing has taken off because carsharing, unlike sharing smaller bikes and e-scooters, is logistically challenging. Cars travel further away from each other and are bigger and heavier to move so that rebalancing the fleet to satisfy real-time demand is far more difficult and costly. Ride-hailing (e.g., Uber, Lyft) provides similar door-to-door convenience as carsharing but without the costly investment in vehicles, insurance, refueling or recharging, and fleet rebalancing that carsharing operators, such as Zipcar and Car2go, have faced (because the ride-hailing driver takes care of these tasks). Unlike carsharing, the user does not need a driver's license to hail an Uber.

Micromobility solutions, such as bicycle-sharing (e.g., Citi Bike, Mobike), provide a convenient, affordable, healthy, and eco-friendly mode of transportation for typical distances of a mile or so (**Figure 3.18**). Typical users are young and agile, and are often more tolerant of mixing with other vehicular traffic, despite safety concerns. Micromobility services function as a first- and last-mile solution and can often provide a faster alternative to walking or ride-hailing. As an example, Glyde is an Indian startup that is beginning to create an on-demand, app-driven electric scooter rental program in Mumbai to address the last-mile connectivity with the bus and train system in a more affordable manner than hailing a tuk-tuk or taxi and offering a more comfortable experience than walking [3.22].

Figure 3.18 Some micromobility choices (as seen in Budapest and Amsterdam, respectively).

However, micromobility in general has several issues, such as creating clutter on sidewalks, and safety concerns regarding sharing road space with vehicles or sharing sidewalks with pedestrians. As with carsharing, micromobility has faced profitability challenges because the operational costs are high due to a frequent need for vehicle rebalancing, repairs, recharging and replacement, given the heavy use, theft, and vandalism. An analysis by Quartz of publicly available trip-level e-scooter sharing data from Louisville, Kentucky concluded that a Bird e-scooter only lasted for about a month, and with an average of under four rides per vehicle per day, Bird lost nearly $300 on each e-scooter [3.23].

Micromobility usage also drops off during bad weather and in winter, and there may be market saturation in the city leading to price wars.

In addition to upfront costs for the vehicles, operating license, and app, there are also charges imposed by the city for dock-less vehicles. Safety requirements and insurance costs, because of accidents and injuries, must also be covered. While these services have gained popularity and have shown potential for sustainable urban mobility, achieving profitability has been challenging. Some companies are exploring new business models, partnerships with cities, or integration with public transport to improve their financial sustainability.

Micromobility technology will continue to evolve with new microvehicle designs for a variety of traveler needs, such as microcars for weather protection and even sit-down e-scooters for travelers who do not want to stand up. The durability of microvehicles will improve as next-generation models are developed for shared, commercial use instead of using existing consumer-grade scooters, for example [3.24]. These "optimized" scooters, such as the Bird Three, have 1 kWh batteries that double the range so that they do not have to charge as frequently, and they have also been designed to be more difficult to steal. Lime, meanwhile, promotes battery swapping so that gig workers can recharge the batteries, and this avoids vehicles needing to be collected, transported to charging facilities, recharged and redistributed.

MaaS integrates micromobility with public transport and other mobility services, often provided by different companies, into a single platform or app to provide personalized, on-demand mobility solutions. The more successful MaaS apps allow users to access different forms of transport (buses, trams, metros, trains, city bikes, e-scooters, taxis, and car rentals) with a subscription or pay-as-you-go model. Routes can be planned and tickets can be purchased for different types of transport, which makes switching between modes during a single journey far easier. The aim is to free people from having to own a car in order to have a good life (**Figure 3.19**).

In short, MaaS is being driven by several trends that are unlikely to reverse. These include digital innovations such as smartphones, cloud computing, seamless payments, and data analytics as well as societal trends like urbanization, sustainability, and public-private partnerships.

Figure 3.19 A typical example of a MaaS app.

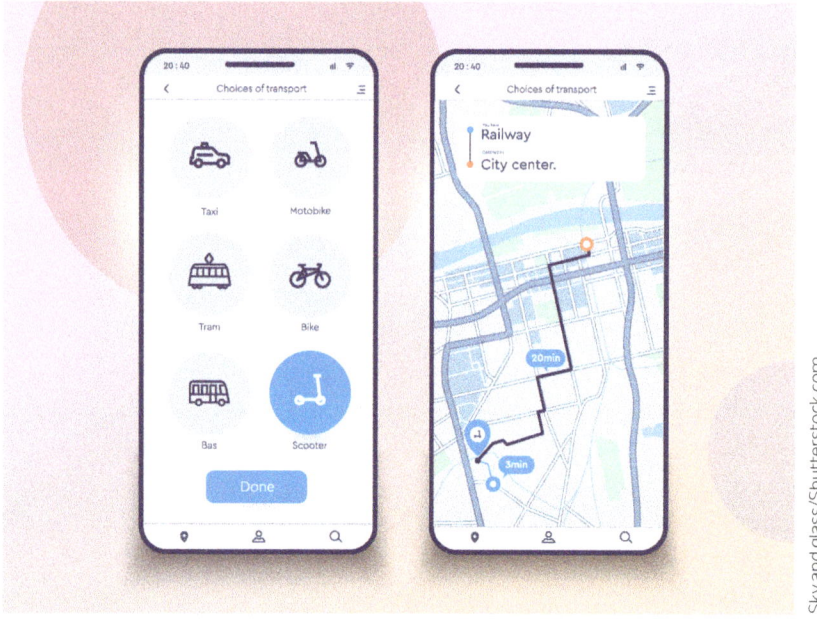

MaaS tends to operate on a city basis where multimodal transport choices exist and where private car usage is most challenged, because of parking and congestion. However, there is nothing inherent to MaaS that limits it to urban areas, and it could be expanded if it integrates peer-to-peer carsharing, for example. Car owners can be paid to rent out their idle vehicles and users have access to a diverse range of vehicles at a lower price than is usual with car rental. As with vehicle subscriptions, peer-to-peer carsharing offers flexible access to vehicles without the burdens of ownership, appealing to those who may not need a car daily. The main challenges with adoption are developing trust between private owners and users, insurance liability issues, cleaning and refueling, and regulatory uncertainties. These challenges and potential solutions will be discussed more extensively in Chapter 7, Section 7.4.

Some African Innovation

In several visits to Africa, I have seen firsthand the resourcefulness and innovation being applied to create new solutions, automotive and otherwise. At the COP22 event in Marrakesh, in November 2016, one Moroccan inventor showed me his 3D printer. He had built it from scrap yard waste and was also using scrap plastic bottles to make the plastic filament material used by the 3D printer. Can waste plastic be "printed" into low-speed EV parts if performance requirements for strength and

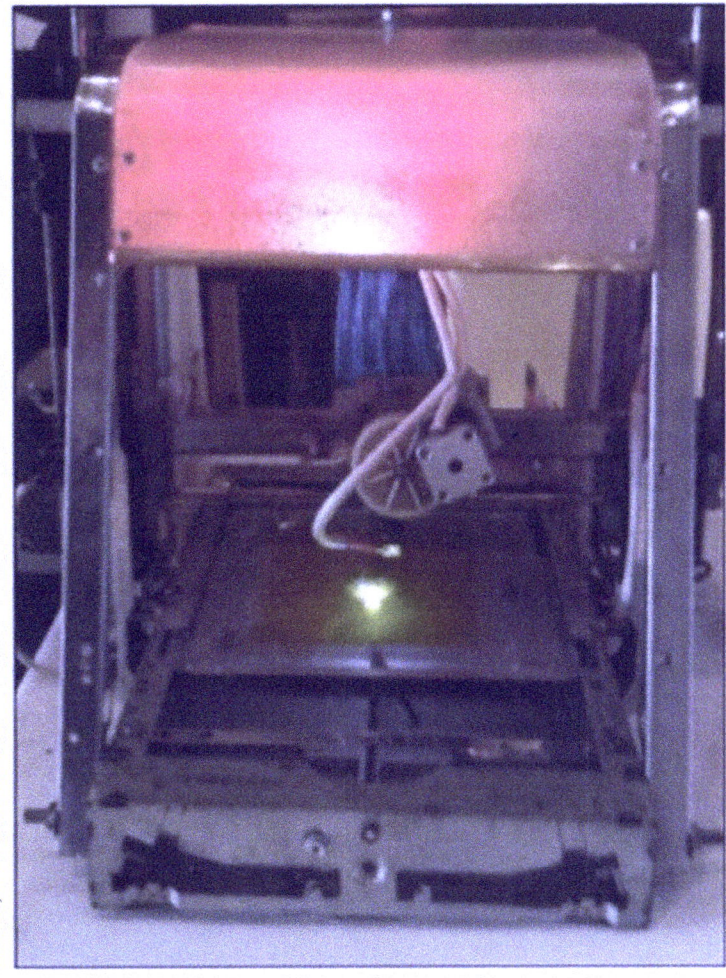

A 3D printer made from recycled plastic parts.

stiffness can be relaxed because the vehicles travel at low speeds and do not have to be designed to meet automotive crash standards? This will be discussed in Chapter 9, Section 9.2.

At the same event, I heard a story of how an East African team had attached a cell phone to a drone so that when it was flying above a forest the microphone would "wake up" when it heard the characteristic sound of a chain saw (illegally chopping trees down). It would then photograph the poachers, for the purpose of identifying and publicly shaming them, and would drop seeds over the deforested area so that trees could grow back again in time. This integration of a mass-produced phone onto an "EV" could be extended to using the smartphone's cameras, GPS, and compute platform to enable affordable, low-speed autonomous operation.

With Lincoln Wamae (2nd from right), a Nairobi-based inventor of electric wheelchairs that reuse laptop batteries.

In March 2020, just before COVID-19 shutdowns occurred worldwide, I was in Zimbabwe and Kenya with a colleague, Richard Saad. In Kenya, we met an entrepreneur, Lincoln Wamae, who scavenged for used laptop batteries and would make new battery modules from them that could electrify a wheelchair and make it easier for people with disabilities to move around [3.25, 3.26]. Repurposing used lithium-ion battery cells, or modules, for low-speed EVs is a logical development because used cells typically have reduced performance, even though this is still more than adequate for low-speed EVs. The e-kit, described in Chapter 8, Section 8.6.2, takes this approach.

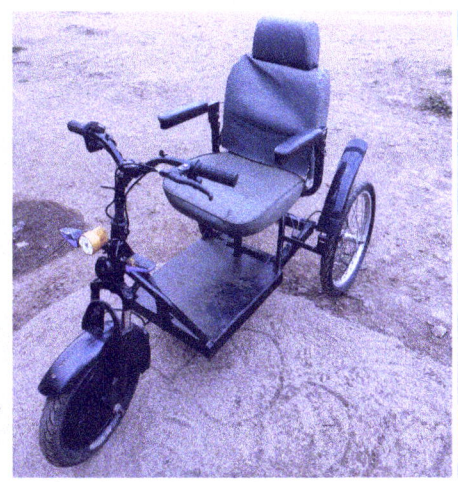 

Electric wheelchair with reused laptop batteries inside the floor.

Zambikes, a Zambia-based Company, has made bicycle frames from bamboo [3.27, 3.28]. Locally available and renewable materials could be used to make low-speed EV structures, improving sustainability while creating jobs. This theme will be covered in Chapter 9, Section 9.3.

References

3.1. Reuters, "Berkshire Hathaway Accelerates Sales of China's BYD," June 25, 2024, https://www.reuters.com/business/autos-transportation/berkshire-hathaway-accelerates-sales-chinas-byd-2024-06-25/.

3.2. Autovolt Website, "New Tech in Nissan Leaf Diagram," https://www.autovolt-magazine.com/new-tech-in-nissan-leaf-01/.

3.3. Palmeron, C., "Rwanda's Shift to Electric Motorcycles Brings More than Just Climate Benefits," How We Made It in Africa, May 15, 2024, https://www.howwemadeitinafrica.com/rwandas-shift-to-electric-motorcycles-brings-more-than-just-climate-benefits/170646/.

3.4. The Economist, "Africa's EV Revolution Has Two Wheels Not Four," October 19–25, 2024, 40, https://www.economist.com/middle-east-and-africa/2024/10/17/africas-ev-revolution-has-two-wheels-not-four.

3.5. Conzade, J., Engel, H., Kendall, A., and Pais, G., "Power to Move: Accelerating the Electric Transport Transition in Sub-Saharan Africa," McKinsey, February 23, 2022, https://www.mckinsey.com/industries/automotive-and-assembly/our-insights/power-to-move-accelerating-the-electric-transport-transition-in-sub-saharan-africa.

3.6. Volta Foundation, "2024 Battery Report," 59, https://volta.foundation/battery-report-2024.

3.7. Battery Design from Chemistry to Pack, "NMC vs LFP Costs," December 10, 2024, https://www.batterydesign.net/nmc-vs-lfp-costs/.

3.8. Frith, J.T. et al., "A Non-academic Perspective on the Future of Lithium-Based Batteries," *Nature Communications* 14 (2023): 420, doi:https://doi.org/10.1038/s41467-023-35933-2.

3.9. Zheng, R. et al., "State of the Art of Lithium-Ion Battery SOC Estimation for Electrical Vehicles," *Energies* 11, no. 7 (2018): 1820, doi:https://doi.org/10.3390/en11071820.

3.10. Anderson, B., "Nio CEO Drives ET7 for 650 Miles on a Single Charge," CarScoops.com, December 19, 2023, https://www.carscoops.com/2023/12/nio-ceo-drives-et7-for-650-miles-on-a-single-charge/.

3.11. Binns, T., "Samsung Shows Off New Solid State EV Battery, with Over 600 Miles Range," Electric Drives, July 29, 2024, https://electricdrives.tv/samsung-shows-off-new-solid-state-ev-battery-with-over-600-miles-range/.

3.12. Reuters, "Why Are Chinese Automakers Like BYD Launching Fast-Charging EV Systems?," March 18, 2025, https://www.reuters.com/business/autos-transportation/why-are-chinese-automakers-like-byd-launching-fast-charging-ev-systems-2025-03-18/.

3.13. Zhang, P., "Most of Nio Swap Stations in Shanghai Deliver over 100 Services per Day, Says CFO," CNEV Post, November 14, 2024, https://cnevpost.com/2024/11/14/most-nio-swap-stations-shanghai-over-100-services-per-day/.

3.14. CATL Press Release, "CATL Launches Battery Swap Ecosystem with Nearly 100 Partners," December 18, 2024, https://www.catl.com/en/news/6342.html.

3.15. Fowler, G.A., "What Does Your Car Know about You? We Hacked a Chevy to Find Out," Washington Post, December 17, 2019, https://www.washingtonpost.com/technology/2019/12/17/what-does-your-car-know-about-you-we-hacked-chevy-find-out/.

3.16. Billington, J., "The Prometheus Project: The Story behind One of AV's Greatest Developments," August 22, 2018, https://www.autonomousvehicleinternational.com/features/the-prometheus-project.html.

3.17. USDOT FHWA "Demo '97: Proving AHS works," July/August 1997, Vol. 61, issue no.1, https://highways.dot.gov/public-roads/julyaugust-1997/demo-97-proving-ahs-works.

3.18. Hinton, G. et al., "Deep Neural Networks for Acoustic Modeling in Speech Recognition," *IEEE Signal Processing Magazine* 29, no. 6 (2012): 82-97, doi:10.1109/MSP.2012.2205597.

3.19. "Driving NYC Taxis Out of Business: How Uber and Lyft Doomed the Once-Solid Yellow Cab," *New York Daily News*, January 30, 2020, https://www.nydailynews.com/2020/01/30/driving-nyc-taxis-out-of-business-how-uber-and-lyft-doomed-the-once-solid-yellow-cab-industry/.

3.20. Salam, E., "'They Stole from Us': The New York Taxi Drivers Mired in Debt Over Medallions," *Guardian* (2021). https://www.theguardian.com/us-news/2021/oct/02/new-york-city-taxi-medallion-drivers-debt.

3.21. Bursa, M., "London Black Cab Could Face Extinction by 2045, According to New Report," Professional Driver Website, March 19, 2025, https://www.prodrivermags.com/news/london-black-cab-could-face-extinction-by-2045-according-to-new-report/.

3.22. ANI, "Can Glyde Solve Urban Mobility's Biggest Challenge?," March 22, 2025, https://www.aninews.in/news/business/can-glyde-solve-urban-mobilitys-biggest-challenge20250321114924/.

3.23. Griswold, A., "Shared Scooters Don't Last Long," Quartz, March 1, 2019, https://qz.com/1561654/how-long-does-a-scooter-last-less-than-a-month-louisville-data-suggests.

3.24. Hawkins, A.J., "Bird Has a New Electric Scooter: It's Durable, Comes in Three Different Colors, and You Can Buy It," The Verge, 2019, https://www.theverge.com/2019/5/8/18535698/bird-one-electric-scooter-ride-share-own-price.

3.25. Kamau, H., "Innovation Spotlight: From Junk to Wheelchairs, the Story of Lincoln Wamae – An Engineer and Innovator," Global Disability Innovation Hub Website, https://www.disabilityinnovation.com/blog/from-junk-to-wheelchairs.

3.26. Muigai, N., "Kenyan Innovator Converting Trash into Electric Wheelchairs," September 2, 2019, https://www.bbc.com/news/av/world-africa-49528426.

3.27. "Zambikes: Bamboo Bikes that Help Fight Poverty & Save Lives in Africa," Slideshow, https://inhabitat.com/zambikes-bamboo-bikes-that-help-fight-poverty-save-lives-in-africa/zambikes_team/.

3.28. The Habari Network Website, "Zambikes: Bamboo Bicycles Made in Zambia," https://www.thehabarinetwork.com/zambikes-bamboo-bicycles-made-in-zambia.

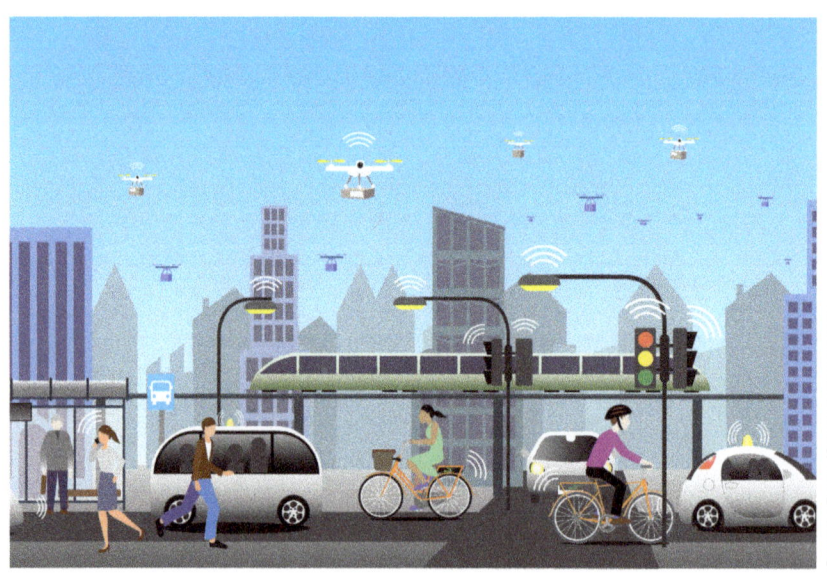

Chapter 04

The Emerging Automotive Landscape 2025–2035

This chapter will discuss some of the emerging technology solutions that are currently being developed and may soon see market introduction. It is not intended to be an exhaustive list but is meant to build on the recent developments that were discussed in Chapter 3, with a focus on electrification, connectivity, and autonomy.

4.1. Battery Performance

Over the next ten years, significant improvements are expected in battery cost, performance (energy and power), and lifespan, which will address the current limitations of EVs and make them competitive with gasoline ICEVs. An example of a battery technology roadmap is shown in **Figure 4.1**, which highlights projected progress in battery chemistry in the near term. As discussed in Chapter 3, Section 3.3, sodium-ion battery chemistry is emerging as a way to reduce the environmental and supply chain security concerns associated with the raw battery materials, as well as to improve inherent battery safety without needing as much active monitoring. CATL expects that when its current sodium-ion battery is scaled for mass production, the manufacturing costs will be much lower than the current most affordable battery chemistry (LFP) while the specific energy, 200 Wh/kg, is already approaching that of LFP. This could make it the logical choice for micromobility and mass-market EVs as well as for renewable energy storage [4.1]. With appropriate thermal management, ultrafast charging rates exceeding 1000 kW are possible, significantly reducing charging times and making them comparable with the time it currently takes to refuel with gasoline. Thus, the price, range, and refueling time barriers to EV adoption should fall.

Figure 4.1 Battery technology roadmap.

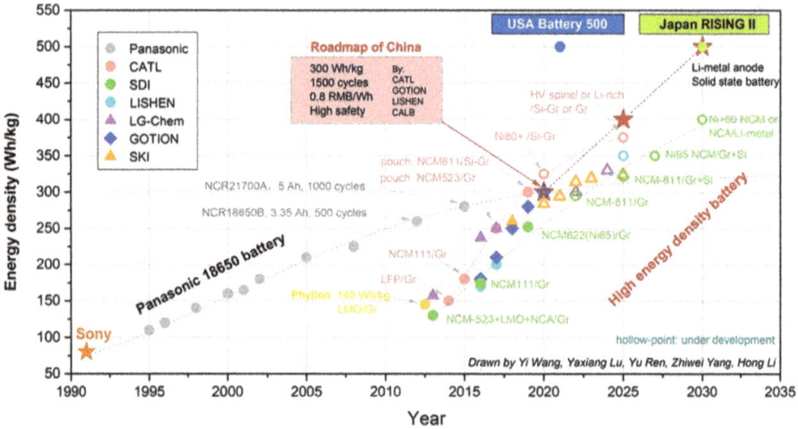

Used with permission of Elsevier, from "Research and development of advanced battery materials in China," Lu et al., Volume 23, 2019; permission conveyed through Copyright Clearance Center, Inc.

In terms of overall vehicle lifetime ownership costs, beyond initial purchase cost, consideration must also be given to "fuel" or energy, maintenance, and repair costs. Compared to ICEVs, EVs benefit from lower energy cost per mile driven (roughly 1/3 for overnight, residential charging that uses off-peak electricity), because the average EV sold in the US has an equivalent fuel economy of 91 mpg. EVs should need less maintenance because there is no need for regular oil changes [4.2]. In addition, engines and transmissions have many moving and nonmoving parts that can fail, whereas electric motors are simpler and have fewer components that can break. Friction brakes should also last much longer because EVs (as well as other types of EVs, such as HEVs, PHEVs, and FCEVs) tend to slow down by feeding energy back into the battery, rather than by immediately applying friction brakes as ICEVs do. Mercedes-Benz has claimed, in fact, that 98% of the braking on its EVs is regenerative [4.3]. If the EV is much heavier and is accelerated more aggressively, then the tire tread may wear out sooner, but with battery improvements making weight less of an issue and with appropriate driving, it is likely that future EV tires could have a similar life. Moreover, tire suppliers are researching ways to reduce tire wear as part of the Tire Industry Project [4.4], and they are developing tires that are optimized for EV usage, including lower rolling resistance. Although data from the J.D. Power 2024 US Initial Quality Survey indicates that EVs have more problems than ICEVs, a closer study reveals that the issues are mainly with advanced software and infotainment features, which are typically more common in higher-tech EVs than in ICEVs [4.5]. In 2025 Germany's roadside assistance organization (ADAC) announced that after responding to 3.6 million breakdowns in 2024 they found that EVs averaged 4.2 breakdowns per 1000 vehicles while comparable ICEVs had 10.4 breakdowns per 1000 vehicles [4.6].

What has been a more significant concern for consumers is whether the battery life is comparable to that of an ICE propulsion system (including the transmission), which is around 100,000 miles. Extensive data analyses of how vehicles are being used, together with improvements in battery materials, battery management, and thermal management systems that maintain optimal battery temperatures are all helping to boost battery life to the point where cycle life (number of times the battery can be charged and discharged while retaining at least 80% of the battery's initial capacity) may make it possible to operate the vehicle for over half a million miles! This would make them very attractive to fleets, particularly for AV fleets that drive nearly all day without needing a rest stop for a human driver, and could drive well over 100,000 miles a year. Of course, for passenger vehicle manufacturers, the same improvement could lead to future customers keeping hold of their vehicle longer, which would reduce new vehicle sales!

Safety is always a concern with any source of substantial energy. Gasoline can catch fire, but because a battery fire can be faster to propagate, longer lasting, and harder to extinguish, it is imperative to ensure safe battery operation and protection. Improved BMSs, which leverage machine learning and are augmented by chemical or thermal sensors, can monitor and manage battery conditions and prevent safety issues from arising in the first place. These same systems also extend battery life. Moreover, future sodium-ion or solid-state lithium-ion battery technology is expected to significantly improve battery safety by further reducing the risk of thermal runaway and fires in the first place.

4.2.
Battery Charging

With battery innovations that address vehicle cost and driving range, the other main customer pain point is with charging EVs. Even though most passenger vehicles drive around 30 miles per day, on average, and EVs can be recharged overnight, there are many people, such as apartment dwellers, who do not have easy access to charging at home and need to find public charging stations. There are also times when an EV owner wants to take a longer trip and needs to find not only a public charging station but a fast one as well. People are used to taking short breaks to refuel their cars, and EV charging should not take any longer. To fully recharge a 300-mile range EV that has a 100 kWh battery pack requires recharging at a rate of 1200 kW if it is to happen in five minutes (by comparison, Level 1 charging is usually no more than 3 kW and even installed Level 2 residential charging is typically only around 10 kW). Batteries have been developed to accept higher Level 3 DC fast charging rates without significant degradation, even if such fast charging is done daily, as it might for fleet vehicles that drive 300 or more miles every day. This type of ultrafast charging has been enabled by the development of high-voltage power electronics and advanced cooling systems. As mentioned in Chapter 3, Section 3.3, batteries made by CATL and BYD are beginning to meet these ultra-fast charging rates without reducing battery lifespan.

Although this type of Level 3 DC ultra-fast charging, faster even than the Tesla Supercharger, may be needed to address range anxiety, there are several other approaches to recharging the battery that could be more attractive in certain situations even if they are slower. One of these is wireless charging, familiar to many for charging smartphones, that can be applied to EVs if the vehicle is equipped with a charging pad on the underside of the body and parks directly over a charging pad resting on the ground surface, as shown in **Figure 4.2**. Initial deployment could be with a vehicle that can autonomously park in a residential garage. This would save the driver time and ensure charging always occurs when the vehicle is parked even if the driver "forgets," so that it is charged as much as possible when driven off. Ultimately, the charging pad could be flush-mounted with a street surface, analogous to a manhole cover, or even buried slightly below the street surface so that it is invisible. This has the benefit of reducing cable clutter and protecting the pad from vandalism or abuse. It may also reduce theft of charging cables because of copper's high value.

Figure 4.2 Wireless EV charging.

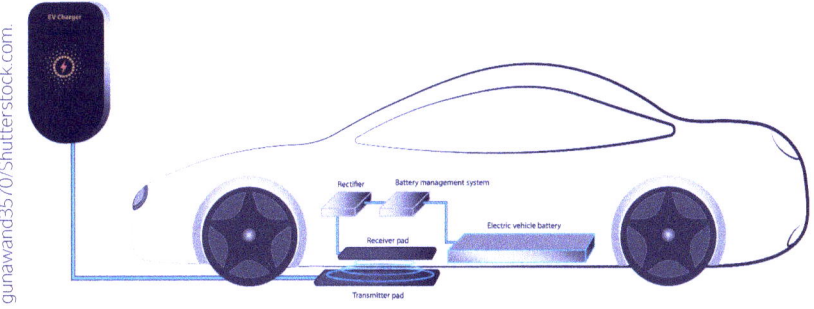

By embedding the wireless charging pad in the road, there is also potential for opportunistic charging when the vehicle is stopped at a traffic light or bus stop, for example. This will extend the EV's range and could even enable a smaller battery to meet route or daily driving requirements for fleets having a well-defined drive cycle. Alternatively, a roadway could have wireless charging pads embedded in it and provide continuous charging for a bus. This concept has been trialed in several locations around the world, including a Korean Advanced

Institute of Science and Technology's electric bus in Seoul's Grand Park since 2013 and, recently, in a one mile stretch of road in Detroit by Electreon (an Israeli company). The main disadvantages of wireless charging are the extra cost for the charging pad in or on the ground as well as on the vehicle, and concerns about how much power can be transmitted across the wireless gap because stray electromagnetic emissions may pose some health risks to people walking by the vehicle or to animals that may go under the vehicle. Some of these risks can be mitigated if the charging takes place in a closed-off area, such as a fleet parking lot.

Automated charging is similar in some respects to wireless charging, because it also makes the charging process more convenient and hands-free. It works by using a robot or automated mechanism to locate and connect the charging cable to the EV's charging port, eliminating the need for manual intervention, as shown in **Figure 4.3**. In the future, the connection could be underneath the EV and would physically dock onto it once the vehicle is precisely parked over it. This would make it even more like wireless charging and could be particularly useful in commercial or fleet settings, where vehicles need to be charged quickly and there may be limited space for parking. The advantage of this approach over wireless charging is that the physical connection could transmit higher levels of power with higher efficiency and without the concerns for radiated wireless emissions.

Figure 4.3 Automated, or robotic, EV charging.

Another aspect of charging is how it integrates with the electric grid and with renewable energy storage. As an example, solar power can be generated and then used to recharge EVs when parked during the day. California is currently a net exporter of solar power to neighboring states because there are not enough EV consumers to recharge during the day, but in the future, given enough EVs, daytime charging from renewable sources might even become less expensive than nighttime charging. In addition to solar-powered charging stations or parking structures, solar charging could occur by integrating solar panels directly onto the vehicle's body. While solar charging alone may not be sufficient for long-distance travel, it can extend the range of EVs and reduce dependence on the grid, particularly on hot summer days when the grid is challenged with supporting HVAC demand. However, for right-sized vehicles in sunny climates, solar charging could provide the majority of energy for the EV, as will be covered in Chapter 8, Section 8.3.

EVs can be used to power electrical appliances and tools with a feature called V2L (vehicle-to-load, where the EV has its own 120/220 V outlets). Automakers are now enabling EVs that accept bidirectional charging so that they can act as energy sources and feed power back to the grid (vehicle-to-grid, or V2G) or to the home (V2H), if the home charger is modified accordingly (**Figure 4.4**). The V2G capability may be included in the automaker's suggested retail price (MSRP), but a 10 kW (Level 2) DC bidirectional charging unit can cost over $1000 presently and needs to be paid either by the charging station or by the individual EV owner [4.7].

Figure 4.4 A Nissan Leaf with V2H capability.

This V2G feature might be mandated in the future by governments because it can help during a natural disaster or when there is an electric grid brownout. Such V2G capability would allow EVs to supply power back during peak electricity demand, contributing to grid stability and reducing household electricity costs. In this vision, EVs act as energy storage units for intermittent renewables, helping to increase adoption of renewable energy. The EV owner might also be able to make money from selling electricity back at a higher price than when it was being charged (**Figure 4.5**).

If an EV is equipped with bidirectional charging capability, as shown in **Figure 4.6**, then it can also recharge another suitably equipped EV when the other EV has a low battery charge and needs more range to reach the nearest charging station, for example. This is V2V charging, and payment systems will likely be developed that allow EV owners to transfer and sell electricity to another EV owner.

Figure 4.5 EVs will be integrated into a new electric mobility ecosystem with renewable power generation and battery energy storage.

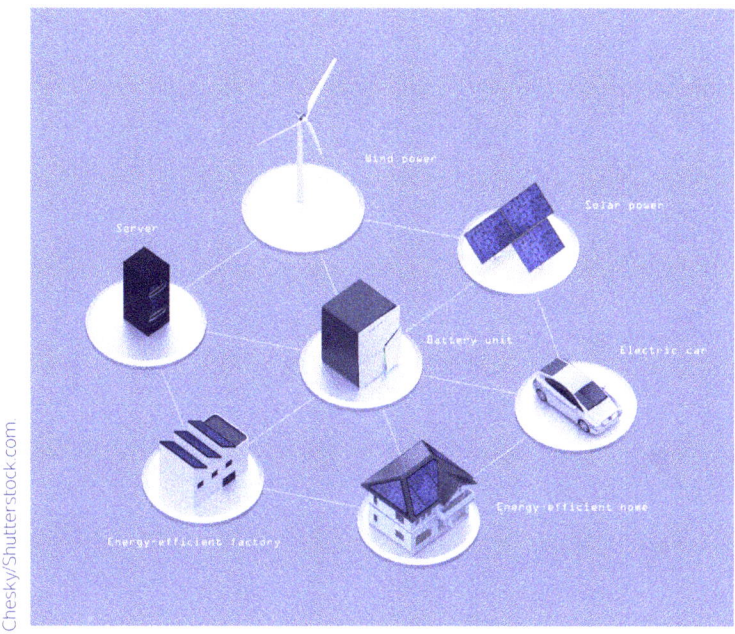

Chesky/Shutterstock.com

Figure 4.6 V2V charging, as shown between a Ford Lightning and a Rivian R1T.

Courtesy of Sam Abuelsamid.

Because it can take several years to upgrade the EV charging infrastructure, V2V charging could play a useful role in locations such as apartment complexes that lack charging stations. When combined with autonomous operation, the EVs might also be able to approach each other, but humans might still be needed to connect the two EVs together!

For EV charging "deserts," there may also be a good business case for a portable hydrogen fuel cell charging station. This could become a temporary electric charging station to recharge EVs in an apartment complex, industrial park, construction worksite, or along a remote stretch of highway. GM's Hydrotec fuel cell can deliver 150 kW DC power and can charge two vehicles at a time, while the system stores 40 kg of compressed hydrogen, which may be enough for a handful of FCEVs. In a manner similar to the way that gasoline is transported to gas stations by truck today, hydrogen can be brought in by truck on a regular basis and the fuel cell system consumes it to generate high-power DC electricity for recharging purposes. This solution can store significantly more energy and deliver significantly more power than a solar panel-battery solution for a given space of land, but the capital equipment and generation costs (especially with green hydrogen) are higher.

These innovations aim to address various challenges associated with EV charging, such as charging times, convenience, range anxiety, infrastructure availability, and emergency charging situations. While some technologies like fast DC charging and battery swapping (Chapter 3, Section 3.3), or solar charging are already being adopted, others like automated charging, wireless charging, V2V charging, and mobile hydrogen refueling are still in the development or early deployment stages.

The future of EV charging is likely to involve a combination of these technologies, tailored to specific use cases and adapting to the evolving needs of EV owners and the transportation industry. As with battery innovation, EV charging innovation means that we are likely to have many more choices in the future for the type of "refueling" experience than we have today with ICEVs!

4.3. Hydrogen Fuel Cells

FCEVs use fuel cells, specifically a proton exchange membrane fuel cell (PEMFC), to generate electricity from hydrogen (fuel) and air, producing water vapor as the only emission. Like battery EVs, they are smooth and quiet, emit no carbon dioxide or harmful exhaust gases, and are also classified as zero-emissions vehicles (ZEVs). One attractive feature of FCEVs is fast refueling, comparable to a gasoline or diesel vehicle, so that it only takes about five minutes to add 400 miles of range. However, the limited availability of hydrogen fuel means that a hydrogen refueling infrastructure needs to be developed alongside vehicle commercialization. Fuel cell passenger cars are being developed by Toyota, Hyundai, and BMW with part of the motivation being to change the "rules of the game" since China currently dominates all aspects related to EVs. These companies also promote the vision of helping usher in a hydrogen-based renewable energy economy that benefits from easier hydrogen distribution, via existing, modified natural gas pipelines, and more energy-dense storage than that provided by batteries.

Due to the fact that viable hydrogen refueling is much easier to accomplish with fleet vehicles that have centralized or well-defined refueling locations, a particular focus of hydrogen FCEV development is for heavy-duty trucks that need to be electrified but still need to carry heavy payloads, which would be significantly reduced with the heavy battery packs needed for a battery-electric truck. However, when volume is the constraint, not weight, then the battery solution might be favored. This could be the case if a heavy-duty truck is used for transporting light goods that take up a lot of space but not much weight, such as fresh produce. **Figure 4.7** portrays the trade-off between various types of EVs, in terms of driving range and cargo hauling capability.

Figure 4.7 xEV types for different applications. Such a chart is constantly subject to change, based on the relative state of battery and fuel cell technology versus the ICE, as well as government regulations.

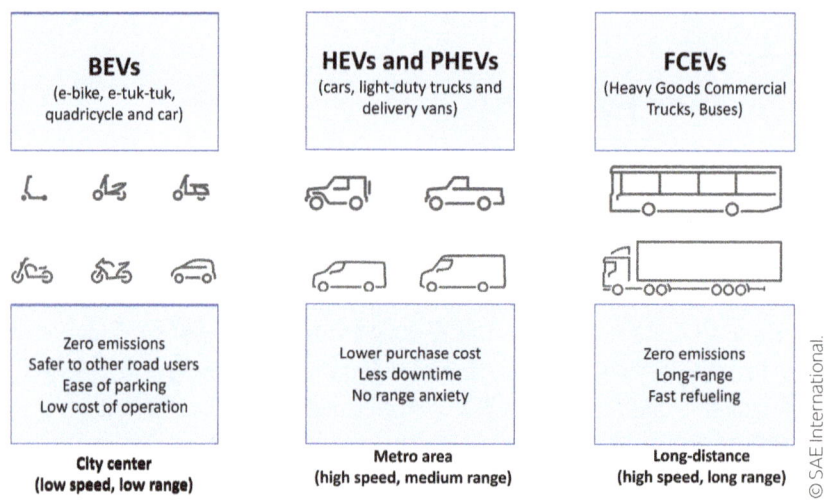

Some of the land vehicle applications where hydrogen fuel cells could make the most sense include long-haul trucks and semitrailers that need high amounts of payload and torque, public transit buses and coaches that must meet zero-emissions targets and have centralized refueling, materials handling equipment (such as forklifts and tow tractors) that operate indoors and can be refueled quickly, and vehicles that are used in construction, mining, and airports that need long-range operation.

The viability of FCEVs as a competitor to BEVs depends on several factors, including advancements in technology, infrastructure development, and market dynamics. Battery technology continues to advance rapidly, spurred by significant research investment, which has meant that previous battery concerns of lower energy density and slower charging are being overcome to the point where the market opportunity for hydrogen fuel cells in land transport applications may be shrinking. However, concerns about the supply chain and geopolitical factors related to battery materials (discussed in Chapter 5, Section 5.3.2) could make FCEVs more appealing to truck makers because they will not need to compete with the automakers for the same battery materials (and likely have lower priority with battery material suppliers because heavy duty trucks represent a smaller addressable market size than passenger vehicles). In addition, the production of hydrogen fuel and of fuel cell systems does not depend on materials that are particularly difficult to find or process.

4.4. Rethinking Chassis and Propulsion

The GM Autonomy concept, described in Chapter 3, Section 3.1, reinvented the automobile's architecture around electrification of all vehicle systems (propulsion, braking, and steering). The concept not only showcased a "skateboard" platform but also included four-wheel motors. These in-wheel or hub electric motors are integrated directly into the wheel assembly (**Figure 4.8**) and can be combined with electric braking and steering systems, suspension, and cooling systems to create a corner module. This design offers several benefits as well as challenges compared to traditional powertrain configurations, electric or otherwise.

Figure 4.8 One example of a wheel motor design.

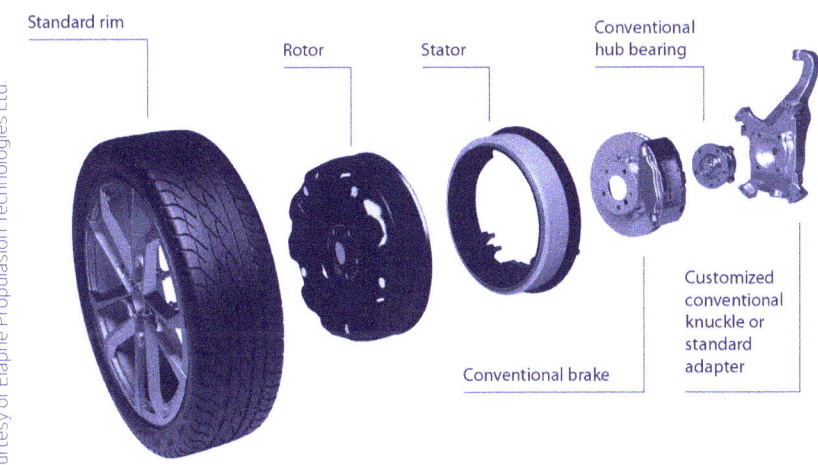

Eliminating the space typically occupied by the transmission, driveshafts, and differential can lead to more spacious interiors and/or a larger "frunk" (front trunk) with the opportunity to improve front crash safety performance. It can also be used to create a more aerodynamic profile, making the solution attractive to both vehicle designers and engineers. From a vehicle dynamics perspective, each wheel can be controlled independently and more precisely, enabling superior torque vectoring and traction control, which leads to better vehicle handling and stability, particularly in challenging driving conditions. The wheel motor corner module is highly modular, allowing for easy integration into different vehicle platforms and configurations, including two-wheel drive, four-wheel drive, or even individual wheel drive setups, as in an E3W. It also means that any repair can be quickly and easily made by swapping the corner module and fixing the problem offline. This productivity enhancement is particularly important for fleets wanting to minimize downtime.

The reasons that wheel motors have yet to be commercialized by the automotive industry are a combination of challenges related to engineering, cost, and even the organizational structure within automakers. The most frequently mentioned technical issue is the additional "unsprung" mass on each wheel that can hurt ride quality and handling, but this can be minimized with properly engineered suspension tuning to a point where only a vehicle dynamics expert can detect any meaningful difference under nearly all driving conditions. Cooling the motor and the associated electronics in the wheel assembly can be challenging in a compact and exposed environment, especially as the nearby brakes can become hot when activated. Although regenerative braking may mean this happens rarely, foundation brakes are still needed for emergencies. The integration of so many vehicle systems (propulsion, braking, steering, suspension, cooling) in a tight space can also increase manufacturing complexity and cost, especially during the early stages of adoption.

Wheel motors do add **motor** cost and mass to the vehicle, and they also need additional cabling and electronics. However, they eliminate several parts of the drivetrain in an all-wheel drive (AWD) vehicle, such as the transmission, driveshaft, transfer case, differential, rear axle, and half shafts. This means that the overall vehicle can become lighter and cheaper when using in-wheel motors, even if the "unsprung" mass is higher. In-wheel motors may also eliminate drivetrain losses, and this may improve energy efficiency by one or two percent, under typical driving conditions, which can translate into more range with the same battery or the same range with a smaller and lighter battery, again saving cost. For low-speed urban vehicles with a low-power motor at each corner, it may even be possible to avoid liquid cooling, relying on simpler, cheaper air cooling because heat removal is no longer centralized and can be distributed at each corner.

Several future mobility trends favor in-wheel motor corner modules. Relatively smooth streets, more goods delivery, and controlled, autonomous operation will reduce the importance of unsprung mass in future urban mobility applications; after all, an AV should be able to detect road bumps and potholes in advance and plan a route that avoids bumpy or poorly maintained roads and the need for tight cornering. Road bumps and potholes may still need to be damped, and this is where an active suspension can provide passengers with an experience that feels more like they are on a train, not on a bus (**Figure 4.9**). An active suspension can also ensure that fragile goods are not damaged in transit and extend the life of vehicle hardware since components experience less "wear and tear." This could lead to fewer repairs and less maintenance on expensive AV hardware, for example. An active suspension system can adjust ride height, making it easier for wheelchair users to enter and exit the vehicle. A complementary "corner" technology—smart tires—can adjust tire pressure, and possibly tread, in anticipation of changing road conditions. This could help improve real-world fuel economy and either prevent a crash or reduce its severity. In short, the e-corner will be a major hardware differentiator for future vehicles.

Figure 4.9 Active suspensions, like wheel motors, are an emerging technology with several advantages for moving goods and people, particularly for fleet vehicles and even more so, for AV fleets.

Increased fleet ownership for shared vehicles and goods delivery will drive decisions based on lifecycle cost, and eliminating the gearbox and driveshafts should make vehicles with in-wheel motor corner modules more reliable and cost-effective over the life of the vehicle. Urban mobility, where a greater fraction of the world's vehicle miles traveled (VMT) are being driven, will place more importance on vehicle compactness, and in-wheel motor corner modules can enable shorter vehicles to have the same occupant and battery storage space, and also support relevant urban performance like "turn on a dime" operation and even sideways translation. For robotaxi operators that need to maintain and store vehicles overnight, more compact and maneuverable vehicles can reduce the amount of land space needed for vehicle storage and allow faster passenger pickup and drop-off times, leading to more paid rides per day and increased profitability. Eliminating the drivetrain can also help to lower the load floor, which not only improves loading/unloading but can improve entry/egress for an increasingly aging population and for wheelchair users. Car-free city centers, discussed in Chapter 9, Section 9.5, are being proposed by several European cities and could stimulate new door systems, such as front entry with a requirement for low step-in height and floor. For all these use cases, the advantages of in-wheel motor corner modules clearly outweigh the challenges, and these early market applications will be crucial for the successful adoption and widespread implementation of this technology approach across even more vehicle segments in the future.

Wheel motors are being developed by a variety of companies, including "pure play" developers, like Elaphe (Slovenia) and Protean (UK), who have both been refining their designs for over a decade. Several automotive suppliers, like NTN and NIDEC (both from Japan), have recently made production announcements and demonstrated vehicle concepts, and startup "skateboard" developers, such as REE (Israel), are integrating wheel motor corner modules in their "ground-up" EV platform solution. As mentioned previously, wheel motor corner modules will not only shape the form and function of future vehicles but, like the skateboard architecture (**Figure 4.10**), will also influence how vehicle manufacturers develop a product portfolio and how they will be manufactured and serviced. Schaeffler (Germany) has envisioned a vehicle portfolio [4.8] that uses a common wheel motor corner module across multiple body types. Independent wheel motor corner modules that integrate braking, steering, suspension, and propulsion with the wheel and tire assembly can make it easier to develop a wider variety of vehicles off the same architecture because it becomes easier to change both track and wheelbase. It can also lead to a "plug and play" philosophy where the whole module is "bolted on" to the skateboard on the assembly line and quickly swapped out when repairs are needed to any part inside the corner module. It is conceivable that a small, medium, and large corner module (with customization of power and torque) could support an entire vehicle portfolio, especially as the software will increasingly become the primary differentiator among vehicles.

Figure 4.10 In-wheel motor corner module and vehicle platform flexibility to adjust both length (enabled by skateboard), and width (enabled by wheel motor corner module).

Most automakers are organized into "silos" for powertrain, chassis, electrical, body structure, interior, and exterior, and so the holistic or "joined-up" thinking that is needed to justify in-wheel motor corner modules is typically absent. Startups that are less "silo-ed" and need competitive differentiation are more open to considering wheel motors. In conclusion, wheel motor commercialization will occur because of its profound advantages, but it will require a combination of innovative engineering as well as choosing viable vehicle and business applications.

4.5. A New Automotive E/E Architecture

As the vehicle's electrical demands have increased over time with electrical content (Chapter 2, Section 2.2), and as future vehicles become increasingly software-driven, there is a growing realization in the automotive industry that the vehicle's electrical architecture needs to fundamentally change to enable this. Historically, electronics content has evolved gradually, and this allowed automakers to incrementally add ECUs as each new feature, such as power seats or lane-keeping assist, was introduced. This is now becoming unsustainable because the complexity of linking as many as 100 ECUs with software and wired connections together to enable new features increases time to market and the risk of subsequent recalls, which may also be safety related. A feature such as AEB, for example, requires that the ADAS sensors need to communicate with both the braking and throttle systems as well as with the infotainment display.

Tesla blazed a trail with OTA software updates to the entire vehicle, not just to the infotainment system, enabling additional range, self-driving capabilities, and even vehicle height modifications. This capability was enabled not only by developing the software in-house but by simplifying the electrical architecture and making it more like the one seen in a PC or smartphone that has a single CPU. Tesla has been able to charge some of its customers up to $15,000 for the full self-driving (FSD) software package, effectively receiving millions of dollars and lots of valuable driving data from its customer base while it is still developing the software to meet the objective of fully autonomous operation. This ability to add monetizable features or functionality via software updates has persuaded many other automakers to embrace this type of software-defined architecture and its enabling smartphone-like electrical architecture (shown in **Figure 4.11**), that has a centralized "brain" that coordinates and sends commands to all the vehicle subsystems.

Figure 4.11 (Left) smartphone E/E architecture (with a CPU), and (right) traditional automotive E/E architecture (with a federation of ECUs).

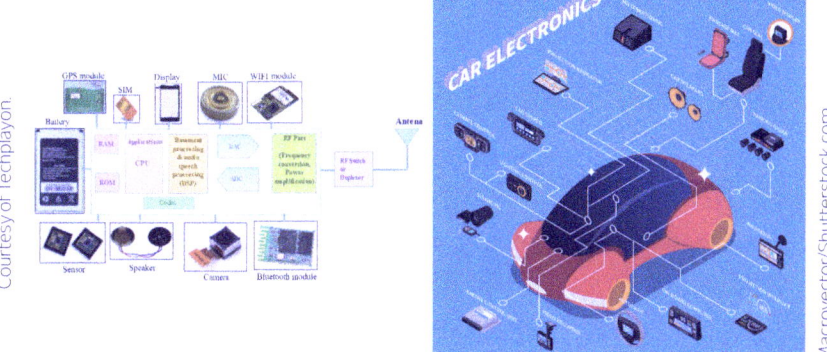

Reinventing the E/E architecture is a disruptive undertaking for traditional automakers, who have many vehicle platforms, powertrains, models, and brands to support while evolving their internal organizational structure and ensuring continuous support to their existing vehicles. Moreover, they need to collaborate with their existing supply base even as this shift threatens to commoditize supplier hardware. Therefore, instead of jumping straight to a single centralized compute platform, automakers are migrating to zonal architectures first, where the vehicle is divided into several zones and each one has a powerful computer, as a transitional step. EV startups, on the other hand, have an advantage because they are focused solely on EVs and can literally start from scratch and copy the Tesla playbook.

4.6.
AVs

The state of the art in passenger vehicle autonomy is currently somewhere between Level 2 (partial autonomy, exemplified by Tesla's Autopilot and GM's Super Cruise) and Level 3 (conditional autonomy, exemplified by Mercedes-Benz's Drive Pilot for highway driving under 60 kph for approved roads and weather conditions) (**Figure 4.12**). In both cases, the system is operational only in specific driving environments (or operating domains), and the driver can take their hands off the steering wheel and feet off the pedals. The difference is that Level 2 (L2) systems require that the driver continues to watch the road ahead, whereas Level 3 (L3) only requires the driver to reengage when leaving the operating domain or if road or weather conditions change significantly inside the operating domain. Another way to think of this is that L2 is "hands-off" the steering wheel, while L3 is "eyes-off" the road as well.

Making the jump from L2 to L3 autonomy involves overcoming some commercial, technical, regulatory, liability, and human factors challenges. For example, to provide superior performance and reliability in all environmental conditions, L3 systems may require the addition of lidar sensing to complement the radars and cameras typically found in L2 systems. L3 also requires more redundancy in computing, power, and communications to provide a higher level of robustness and safety in the

Figure 4.12 Levels of driving automation (or autonomy).

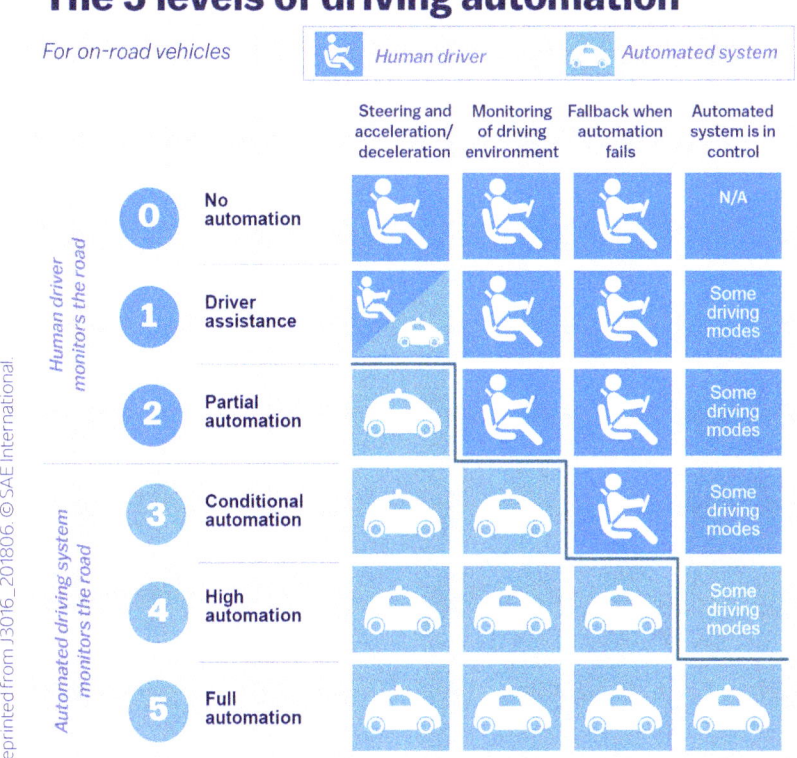

event of any system failures because the automaker is taking responsibility for safely operating the vehicle with less human oversight than for L2 systems where the driver is responsible at all times. A consistent and comprehensive regulatory framework is, therefore, needed to facilitate the deployment of L3 vehicles across various states or neighboring countries, and this means clear guidelines on liability, safety standards, and operational conditions. In 2017, Audi integrated lidar into an Audi A8 sedan and developed an L3-capable vehicle but stopped its development in 2020 because of the lack of a legal framework.

One of the most significant challenges for providing L3 autonomy is to manage the transition of control between the vehicle and the human driver because the latter must take over control when the system requests it, which can be problematic if the driver is not paying attention or is otherwise engaged. This challenge is also present in L2 autonomy because the driver often does not realize that they must always be in control and wrongly assumes that the vehicle is operating with L3 capability. Solutions to the transition of control involve alerting the driver, escalating the intensity of the alert if the driver is not responding, and even bringing the vehicle to a controlled stop on the side of the road if necessary. An L3 system needs to handle, in real-time, rare or unusual situations that it has not seen before or has not been trained on, which means it will need to leverage accurate localization, machine learning, and simulation techniques that are being developed for L4 and L5 AVs (fully autonomous), while also protecting against cyberattacks.

A contentious topic in the AV space is whether Tesla is correct to assume that lidar is unnecessary and if cameras are sufficient for sensing and perceiving the environment. The argument is often made that humans can sense the road environment using eyesight alone. However, cameras are not exactly equivalent to eyes as anyone who has taken a photo probably knows when the picture is not as "good" as what the eyes see. Cameras do not provide the same level of depth and cannot adapt as quickly to changing light conditions. On the other hand, even though our eyes have peripheral vision and we use our ear canal to gain a sense of balance, AVs can have 360° vision and use inertial measuring units to understand pitch, roll, and yaw state, as well as to support knowing the vehicle's absolute position when the GPS signal is lost, so it is difficult to argue that humans have better sensing capability than Tesla's cars.

The bigger problem with using the camera–eye analogy is that it downplays the difference in how perception and prediction are achieved. While the human brain and AV compute platforms share some similarities in their roles (such as processing information and making decisions), they differ significantly in their structure, efficiency, and adaptability. The human brain can perform around one exaflop (10^{18}) operations per second while only consuming roughly 10 W of power, which is approximately 100 times more efficient than the compute platform in an AV, because the brain's complex network of neurons and synapses can achieve extremely high amounts of parallel processing and integrates processing and memory, unlike computers [4.9]. The brain is also much more capable of adapting to and learning from novel or unique situations based on incomplete or ambiguous information, whereas AV compute platforms are specialized, high-powered systems that tend to be designed for specific tasks in controlled environments.

Even when employing advanced machine learning models, they can still struggle with unexpected events that were not part of their training data. In other words, even if computers excel in processing speed and efficiency for specific tasks, the human brain far surpasses computers in terms of storage capacity, adaptability, and complex decision-making. In a sense, AVs lack "common sense" reasoning about human behavior, physical constraints, and social norms that humans take for granted. Understanding that a child chasing a ball might be about to run into the street, or that construction workers might appear unexpectedly from behind equipment requires reasoning capabilities that current AI systems lack. At least for now! This manifests itself in AVs that operate differently than human-driven vehicles in that they may, inexplicably, stop because they have not seen the particular road scenario before and have not been trained on it. I liken this to the famous "Turing" test in that if one were to look down on the traffic from above, then one should not be able to tell which vehicles are being driven by humans and which are AVs or robots. For now, AVs sometimes fail this test.

These unusual situations, or edge cases, are currently the main technical reason why AV technology has not been commercialized at scale. Although AI is universally being applied to solve vehicle autonomy's "long-tail" challenge of rare edge cases, there is a competition between two fundamental software strategies, based on either a deterministic or an end-to-end approach, as contrasted in **Figure 4.13**. Deterministic AI refers to an approach that uses rules, algorithms, and models to make decisions based on a predefined set of inputs and constraints. For autonomous driving, this approach breaks down the driving task into discrete steps or modules, such as perception (e.g., object detection, lane tracking), planning the future path for the vehicle to take, and activating the vehicle brakes, steering, and accelerator functions. Each subtask is handled by a separate software module or component, which is carefully designed and optimized for that specific task. These subtasks are trained with data and leverage neural networks to improve accuracy and performance. The main benefit of this approach is that the decision-making process is transparent and can be debugged more easily in case of a failure. It is also easier to maintain and make incremental improvements. However, as the driving complexity increases so does the number of rules and algorithms, and it can still fail to mimic the performance of nuanced human decision-making.

Figure 4.13 Modular, end-to-end, and hybrid AV software approaches.

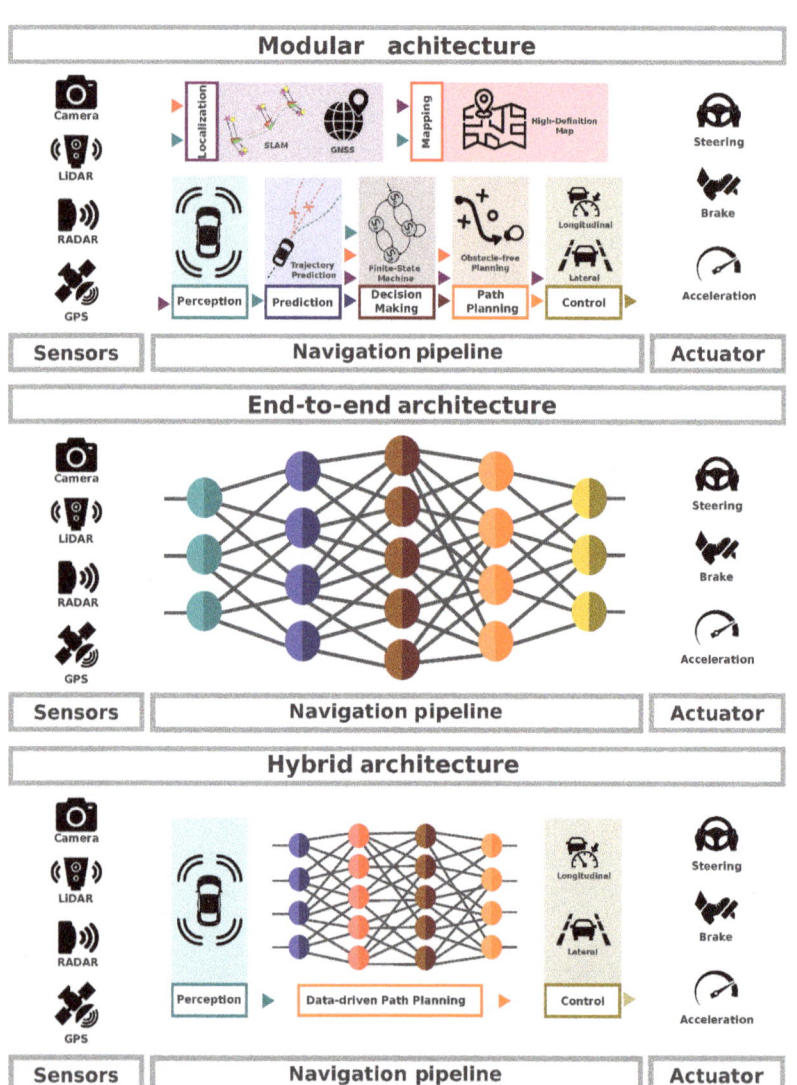

To address this limitation, end-to-end (generative) AI systems learn directly from data, such as sensor inputs and human driving behaviors, without being explicitly programmed with rules or models. These systems are often based on deep learning and neural networks and aim to learn the entire driving task as a single, end-to-end mapping from inputs (such as camera images or radar data) to outputs (e.g., steering, acceleration, braking). In principle, this means that the vehicle could emulate a human-driven one and would not require building a high-definition map before driving on a particular road. It can continuously learn from real-world driving (from its own and from the rest of the vehicle fleet), but it must be able to distinguish and learn from good driving so that it does not learn unsafe driving behavior. It may still struggle with a novel scenario that it has not seen before, even if the frequency of such events is less than with the deterministic approach. A simple example might be that the AV was trained with data to avoid running over a person lying on the road, but struggled with a person who fell onto the road immediately in front of the vehicle. A significant problem with end-to-end systems is their relative lack of transparency, making it hard to understand how the system failed and how to improve it. This also makes it harder to convince regulators of the system's safety.

It is possible that a hybrid approach, combining the best aspects of deterministic and end-to-end AI, may be the most promising path forward. For example, deterministic AI might be used for safety-critical components and well-defined driving tasks, such as highway operation, while end-to-end AI might handle more complex and open-ended aspects of driving, such as decision-making and behavior prediction, and take over in more ambiguous environments, such as residential streets.

It is important to recognize that AVs may always need human support, and so one can argue that they should be called "automated" rather than "autonomous." Even if the software is perfect, an AV can still experience a tire blowout, requiring the vehicle to communicate with the remote operations center (or teleoperations center), and requesting that someone go to the vehicle and replace or repair the tire. Or a passenger may be feeling unwell, in need of medical attention, and driven to a local hospital or caregiver. Or even, in some cases, locking the vehicle and driving it to a local police station. Or an emergency worker may need to communicate with the vehicle to move it to a particular place, rather than have the AV assume where it needs to go. In principle, these problems could be solved "autonomously," but it may be more reliable and economical to use a remote operator, just as one is used today with sidewalk droid delivery robots.

Figure 4.14 shows a teleoperator. How many end up being needed will affect the business case for AV operation because a perceived advantage for robotaxis and autonomous truck fleets is their elimination of labor costs. The business goal will be for as many AVs as possible to be monitored by one teleoperator, and the ratio may need to exceed 10:1 for commercial viability, as well as for user experience (because the fewer times such interventions happen, the better). Ideally, as the teleoperator teaches the AV how to safely navigate through a tricky situation or edge case, the AV learns, the software can be updated, and the solution can be applied across the fleet, so that over time, fewer interventions are needed by the teleoperator, leading to fewer teleoperators being required.

Figure 4.14 Current city traffic management operations may foreshadow a future corporate teleoperations center for privately owned AV fleets.

Remote operators can be the most practical solution for not only solving problems but also creating new business opportunities. For example, they could provide a reliable service helping to move autonomous personal EVs at night to charging stations and returning them to the customer's parking spot in the morning, prior to being used by the customer. This may be particularly attractive for apartment dwellers and in regions, such as China, where homes may not have a charge port and where a remote charging infrastructure can be less expensive to install and have greater utilization to reduce costs. This could be done without a remote operator, but it is likely to be more reliable with their support.

Given the many technical challenges facing AV commercialization, there is a case to be made that the most feasible near-term opportunities are in controlled and semi-controlled environments that have a well-defined operating domain. One example is the short-distance transport of goods from ports to nearby warehouses or rail terminals, where autonomous operation could reduce congestion and address labor shortages. Another could be airports, where AVs could shuttle passengers between terminals and rental car facilities or could streamline baggage handling. A third opportunity might be closed campuses, such as university campuses, business parks, and industrial complexes, that offer a controlled environment that is ideal for the deployment of autonomous shuttles and delivery robots. These "low-hanging fruit" applications are not without challenge, especially if coexistence with people or human-driven vehicles is necessary, and each has a different problem to solve and a different business case, but they are simpler environments to solve and promise improved operational efficiency, cost, and service quality.

Having said that, most of the attention in the AV fleet space is directed toward robotaxis and autonomous trucks because these applications have a much larger revenue opportunity once the technical and regulatory issues are solved. Despite eliminating the driver labor cost, which is roughly half the fare that customers pay, the robotaxi business model is not as clear-cut as it might seem because there are many other costs, some of which are still difficult to estimate or quantify, that Uber or Lyft do not incur but which must be borne by a robotaxi operator.

The most obvious example is the vehicles themselves, outfitted with the autonomy system. Other costs that must be included in the robotaxi business model are the land for parking the vehicles at night, and the infrastructure and labor for recharging, maintaining, and cleaning them. Land is not cheap, particularly in a dense city center, and the cost of the parking spot, perhaps $1000 per square meter, could cost as much as the vehicle itself. If the vehicles are parked at the edge of the city, then the land is less expensive, but the AVs will waste time and miles driving into the city each morning and back to the suburbs at night. The AV cost will go up further if autonomy sensors need to be repaired or replaced during the expected robotaxi life of around 300,000 miles or four years. Vehicle storage and maintenance costs are covered by the ride-hail driver today but will need to be covered by the robotaxi operator. In addition, there are costs associated with insurance, remote operations support, and cellular service, as well as R&D to continually improve the system and protect against cyberattacks. Many think that AV insurance costs will be less than the ~$5000 taxis pay each year because AVs should not crash as often but this is offset by the fact that the cost of repair is likely to be higher for any crash, and there is still a need to insure the software and teleoperations center, for example.

Remote operations, by analogy with air traffic control, are a significant cost burden and there is a strong incentive to increase the number of AVs that can be reliably monitored and supported by each operator to as high as possible and to more than ten, at a minimum. On top of this, if the robotaxis join an existing ride-hailing network to access a wider pool of customers, they may need to pay for this access, potentially adding 20% to their costs. If utilization is around 50%, the cost of operating a robotaxi could approach $3/mile, which may be higher than that for Uber or Lyft. **Table 4.1** attempts to capture the various costs associated with operating a robotaxi service and provides estimates for each, although it must be said that this data is difficult to obtain because robotaxi developers are quite secretive about these numbers. There are significant uncertainties surrounding the numbers, and it can be argued that some of these figures will decrease as the business scales, but many of them may not.

It has not been included here (unless it is absorbed into the license fee) but city transport policy may charge robotaxis a "zombie" fee if they

Table 4.1 Example estimates for robotaxi economics.

	1000 vehicles, 75k miles/year, 300k life (4 years)	
	Cents per mile	
Vehicle	10	No battery replacement, $30k base vehicle cost
Autonomy system	7	$10k + average of one complete replacement per vehicle (damage, vandalism) →$20k
Electricity "fuel"	12	2.5 miles/kWh (300 Wh/mile + 100 Wh/mile compute), $0.30/kWh
Storing vehicles (parking)	6	$400/month/vehicle for covered parking
Maintenance, Cleaning	10	$4k/vehicle each year for labor (~30 minutes/day with $20/hr labor), $2k for replacing tires = $0.08/mile, plus other parts (wear)
EV charging infrastructure	5	50 DC fast chargers (1 charger per 20 vehicles), $300k each ($25k charger, $125-450k installation)
Cellular, Remote Operations	15	One person/20 vehicles, $200k labor cost (like air traffic control) ->$10k/vehicle
Software R&D, Cloud support	10	Estimate, 5-10% of total cost per mile given the high ongoing development and compute costs
Marketing	2	Promotional costs, as with TNCs
Insurance	20	Taxi is $5-10k/vehicle/year (AV should be higher); expensive tech, software failure/hacking, remote operations → $15k/vehicle
Licensing	25	Medallion cost is $12k/year/taxi -> $0.16/mile, but robotaxi will likely be more. Includes wheelchair accessibility requirements.
sum	**$1.22/mile**	
20% network access fee	$1.46/mile	Cost of accessing ride-hailing network
Utilization	$2.44/mile	Robotaxi has a fare-paying passenger inside it for 60% of the vehicle's operation.
FINAL TOTAL COST	**$2.44/mile**	**Need to reduce remote intervention rate and insurance premiums. Need to increase utilization rate and vehicle lifespan.**

contribute to congestion while picking someone up with no one in the vehicle, or they may require that the robotaxi operator serve the disabled or underserved communities, which may require extra vehicle and infrastructure costs for accommodation and be less profitable, as a *quid pro quo* for operating in the lucrative city center. The extra cost to design the robotaxi to accommodate wheelchair users (Chapter 5, Section 5.6.2), for example, could be less expensive than the tax imposed by the city (Chapter 7, Section 7.6).

On the "positive side" for the business case, Alphabet (Waymo's owner) is known for monetizing data, and the Waymo AVs are likely to use the interior camera to determine the rider identities and may share this data for personalized ads, unless the rider remembers to opt out of this data sharing. Monetization of the surveillance economy is, unfortunately, a trend we have to be concerned with and will be covered in Chapter 5, Section 5.4.1.

Even though foundational work in AVs occurred in Europe and the US, and the latter may have a lead in AV software development, it is quite possible that China might be the first country to widely adopt AVs like self-driving cars, robotaxis, and commercial trucks. This is because China's government has been aggressive in promoting and funding AV technology as a strategic priority, setting ambitious targets and providing clear regulatory support for testing and deployment, without the issues caused by different neighboring states or regions having different policies, as is the case in the US. The Chinese government also "shapes" public opinion by limiting negative comments and experiences of AVs on social media and is enabling a cellular V2X infrastructure that can support low-latency communications between AVs and the roadside infrastructure. Beijing and Shanghai have both granted permits to robotaxi developers to pick up passengers and China has many new smart cities that are being built from scratch, which could provide controlled environments that are ideal for the initial commercial deployment of autonomous fleets and robotaxis. In Wuhan, a city of around five million people, more than 500 robotaxis are currently being operated by Baidu. Nationwide, it claims to have provided over 11 million rides since 2019 with over a 1000 robotaxis, comparable figures to Waymo in the US.

Many of China's tech companies can exploit huge amounts of real-world driving data being captured in China to train AI algorithms, and Chinese automakers and EV startups are leveraging their lead in clean-sheet EV architectures, investing heavily in self-driving technology, and racing to commercialize semi-autonomy features in the highly competitive domestic passenger car market. BYD's Seagull, with a price tag below $15,000 will feature a version of its "God's Eye" ADAS that BYD says can drive on highways for 600 miles without driver intervention [4.10]. Whereas the Mercedes-Benz Drive Pilot relies on 13 sensors to enable L3 driving on the highway, Chinese automakers are more focused on "eyes-off" driving in urban centers and are incorporating more than 20 sensors into the vehicle at the design stage to be future-proof [4.11]. According to McKinsey, China could become the world's largest market for AVs. They forecast that in 2040, 40% of new vehicles sold in China could be AVs, and these would account for two-thirds of the PMT [4.12].

4.7.
Cabin Experience

The interior is typically one of the last areas of a vehicle to be "frozen" during development, and it was often the case that if a program went over budget late in a vehicle's development, then skimping on the interior was often the easiest way to save money. Moreover, the top car designers tended to be stylists working on the vehicle's exterior appearance rather than industrial designers who were trying to balance aesthetics with visibility, materials durability, and cabin ergonomics. Although styling is still important, the user experience has now become one of the key battlegrounds for vehicle differentiation and consumer purchase.

Since 2010, it can be argued that the most obvious change to a vehicle's appearance has been in the dashboard (or instrument panel), as shown in **Figure 4.15**. Changes have included a shift from analog to digital displays and the elimination of many buttons. Instead of a radio and a CD player, modern dashboards now feature integrated infotainment systems that provide access to various functions, including navigation, music, and smartphone connectivity. These systems are often displayed on large touchscreens, making them more user-friendly and intuitive. In addition, the instrument panel provides real-time ADAS alerts for LDW and blind-spot monitoring, for example. Ambient lighting is often used to enhance the overall visual appeal of the cabin.

Looking ahead, the dashboard is expected to continue evolving with more integration of artificial intelligence (AI)-powered voice assistants and predictive analytics to provide more accurate and convenient personalization. Augmented reality displays may overlay information, such as navigation directions or real-time data, onto the driver's view of the road,

Figure 4.15 Evolution of the dashboard (2000, 2024, and future).

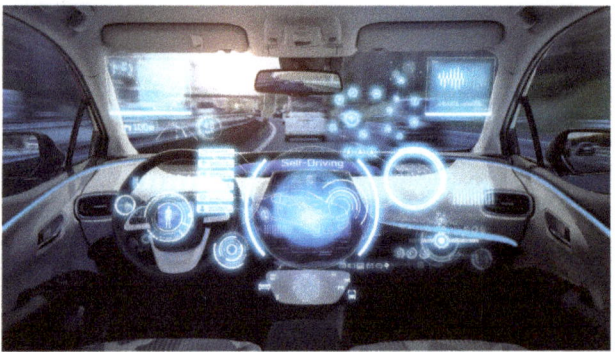

providing a more intuitive experience and minimizing "eyes off the road." In 2025 Hyundai Mobis, in collaboration with Zeiss, introduced a holographic windshield display (HWD) display that is projected onto, and extends the width of, the windshield. It has been designed to allow the driver and front passenger to view different content at the same time without being able to see the other, and the wavelength used means that it is not visible outside the vehicle [4.13]. Meanwhile, Continental is partnering with E-Ink to develop more energy efficient windshield displays that will be increasingly important for EVs [4.14].

Better connectivity with OTA updates will allow for frequent refreshes of software and more tailored, immersive experiences. It should also be possible for an internal digital assistant to replace the function of the instruction manual and to help explain in clear language what is wrong with the vehicle, instead of just providing a "check engine" alert. For example, it could suggest what the driver needs to do in a user-friendly manner and help with other features such as automated parking and infotainment selection. The digital assistant could, in other words, become a virtual friend or front-seat passenger, providing timely, safe driving tips and relevant local fun facts.

AI and connectivity will influence other interior systems as well. We may not realize it, but for more than 25 years, vehicles have incorporated occupant monitoring systems to determine if a seat was occupied or unoccupied, for the purpose of determining whether to deploy airbags in the event of a crash. Since then, more sophisticated systems have been developed to understand the size, weight, and position of the occupants, helping to optimize airbag timing and inflation to minimize injury during a collision.

In 2008, Toyota included a system on their Toyota Crown vehicle to determine if the driver was becoming sleepy by monitoring their eyelids. However, the public's understandable concerns about accuracy and privacy have meant that driver monitoring systems (DMSs), intended to keep the driver alert, have taken longer to commercialize than expected. They have needed a push from regulatory agencies, such as the Euro New Car Assessment Program, to encourage automakers to introduce them, and the EU has recently required drowsiness and attention detection systems for new vehicles. With recent technology improvements other government road safety agencies, including the National Highway Traffic Safety Administration (NHTSA) in the US, are showing interest in a DMS that can reliably detect alcohol impairment, distraction, and drowsiness. It has been estimated by Seeing Machines, a DMS developer, that 12,000 lives could be saved each year in the US if there was a 50% fleet penetration rate for DMS [4.15].

Automakers have recently voluntarily introduced DMS into their semi-autonomous vehicles, such as Tesla's Autopilot and GM's Super Cruise, where it monitors the driver and "prods" them to stay engaged and watch the road ahead even when their hands are not on the steering wheel and their feet are not on the pedals. As mentioned previously, these are both L2 systems whereas the current Mercedes-Benz S-Class has an L3 system that does not require the driver to pay attention. However, the driver still needs to be ready to re-engage when conditions change or when the L3 system reaches the limits of its operating domain (based on speed, type of road, etc.), and so a DMS is required for L3 operation as well.

As with ADAS that fuses information about the external road environment from multiple sensors to get different perspectives and to benefit from their different strengths, it is likely that a robust DMS will need more than just a single camera looking at the driver's face because an inaccurate DMS is likely to become a nuisance and cause the driver to try and disable it or may cause them to purchase their next vehicle from another automaker. Other driver-facing sensors, such as infrared or lidar, and, perhaps, electrocardiogram (EKG) and heart rate sensors that are wearable or embedded in the seat, could be fused together to increase reliability and minimize false positives and false negatives. A camera, for example, can be used to measure the fraction of time the eyes are closed, but when it is combined with heart rate data, it becomes possible to detect more accurately, and perhaps earlier, if the driver is becoming drowsy or if the human "driver" is getting motion sickness after looking down at a smartphone screen for long periods while the vehicle is driving in autonomous mode. Sensing data can also determine if the driver is distracted, stressed, impaired, or having a medical issue. Biometric facial recognition or fingerprint scanning may also be integrated into the dashboard for personalized settings, for authentication to prevent unauthorized use of the vehicle, and for making secure payments.

A logical next step for this increasing "surveillance" is to monitor passengers as well as the driver, perhaps with a central ceiling-mounted camera system, for example. A basic version of this may be used to prevent babies, children or pets from being left unattended in a car and in reducing the risk of heatstroke. More sophisticated solutions, using sensors and cameras, could monitor a person's heart rate, blood glucose level, eye movement, facial expressions, and posture to detect health issues or comfort levels. The cabin environment can also be monitored for temperature, humidity, and air quality. By integrating this information, it should be possible to assess occupant comfort and stress levels and automatically adjust climate, music, light, and seat settings to improve comfort (**Figure 4.16**). If future vehicles have more zonal interiors, with smart seats, then each occupant could have their own customized settings for music, climate control, and adjustable seating. This concept can be taken even further with health and wellness monitoring, especially with an increasingly at-risk, obese, and aging population, and with full AV operation where it is easy to imagine that a person who is too old to drive might be alone in a moving vehicle, which does not happen today.

Because a person spends a significant amount of time in a vehicle, it can also be an effective way to monitor health status over many weeks or even years, leading to more accurate medical diagnoses of chronic conditions. In conclusion, health and wellness monitoring inside vehicles could improve safety, detect early health issues, personalize comfort, and provide wellness insights. However, there are very real privacy, reliability, and accuracy concerns that will need to be managed, as discussed in Chapter 5, Section 5.4.

Figure 4.16 Occupant monitoring for health and wellness.

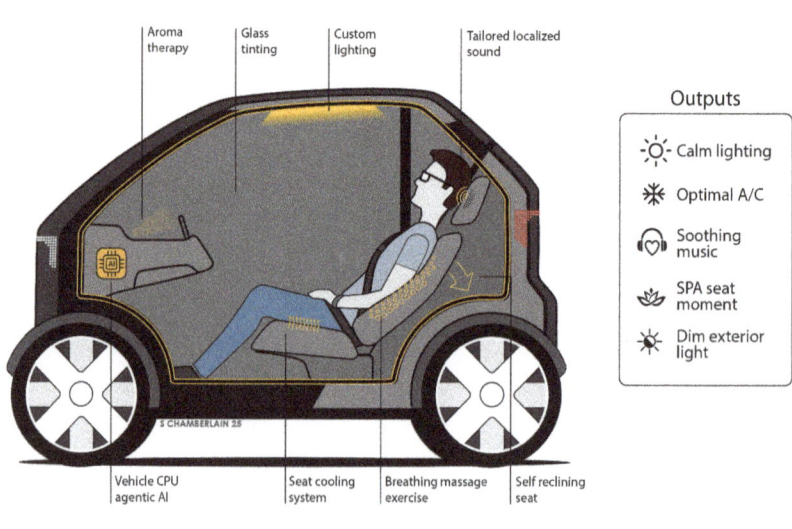

4.8. V2X

AVs, at scale, are likely to increase traffic congestion by inducing more driving miles, but with better coordination, the extra traffic problems can be made less of an issue. That is where V2X comes in because the overall road traffic system can be improved if vehicles can share information with other vehicles on the road, and with the roadside infrastructure (e.g., traffic lights). In one scenario, a vehicle's camera and stability control system engagement might trigger a location-tagged "black ice" condition and relay this information to the cloud so that other vehicles can access this crowdsourced information and take precautions in advance. This should improve road safety and traffic flow, as will traffic lights that automatically adjust red and green light timing based on actual, real-time traffic conditions, with each color change synchronized to adjacent traffic signals to create a "green wave." In the future, it could also help with cooperative traffic management and with coordinated maneuvers for AVs.

The vision of vehicles communicating **directly** with each other (V2V) and with the road infrastructure (vehicle-to-infrastructure [V2I]), without relying on the cloud, is still being researched by automakers, but V2X allows vehicles to directly send and receive messages, sharing relevant information such as location, speed, acceleration, and trajectory. Dedicated radio frequency spectrum has been allocated to ensure that this safety-critical traffic communication is not interfered with by other wireless communications, and many automakers are developing this technology, including Audi, Mercedes-Benz, GM, and Toyota. Since nearly all new vehicles have cellular connectivity, it should not cost much more to add V2V functionality, enabling these vehicles to "see" around corners and in all weather conditions, which will improve the reliability and performance of ADAS. V2V can enable more advanced safety features and more effective forward collision warning, emergency braking, and cooperative ACC because a trailing vehicle can adjust its speed sooner when it receives a direct message from the leading vehicle.

In short, V2V communications would function like a virtual sensor and provide a low-cost method to augment ADAS, further reducing collisions and improving traffic flow. However, it requires that a high number of vehicles are equipped with the same V2V system and that they communicate with each other using the same, standardized communication protocols for seamless operation. Given the collaborative nature of V2V, this will probably require a mandate instead of being left to market forces. This is in contrast with ADAS, which can benefit a vehicle even if other vehicles do not have ADAS!

V2X is the broader, umbrella term that encompasses not only V2V but also other applications that allow vehicles to exchange data with other entities in their environment (**Figure 4.17**). (The V2L, V2G, V2H, and V2V EV charging terms described in Section 4.2 are relevant to EVs only, and are typically not considered in this V2X connected vehicle landscape.) Connected vehicles already connect to external cellular and Wi-Fi networks and can share data with the cloud. This is sometimes abbreviated to V2N (vehicle-to-network), which is also being used to support teleoperation and remote control of AVs, as previously mentioned.

Figure 4.17 Wireless communications for V2X.

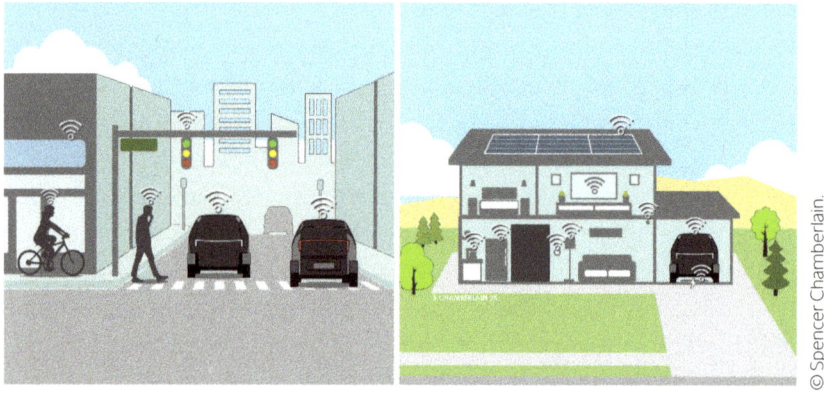

© Spencer Chamberlain.

A distinction can also be drawn between the cloud and the edge. Edge computing refers to processing and analyzing data closer to its source, at the network edge, rather than sending it to a centralized cloud server. For AVs, local data servers may be in city centers or installed alongside the road and provide edge processing to reduce latency and improve real-time decision-making capabilities. This is very important for AVs because they need fast and accurate processing of sensor data to navigate safely and efficiently. Local data servers can also address bandwidth limitations by reducing the amount of data that needs to be transmitted to the cloud for processing, and they can keep sensitive data within the local network, so that AV developers have more control over data access and with a lower risk of data breaches or unauthorized access.

There is also V2I, vehicle-to-pedestrian (V2P), and vehicle-to-cyclist (V2C) communications that can improve safety for vulnerable road users. V2I will let vehicles exchange data with roadside infrastructure, such as traffic signals, road signs, and intelligent transportation systems to provide real-time traffic flow optimization and intelligent parking assistance. In principle, it could also give priority at intersections for high-occupancy vehicles, such as loaded buses, to maximize people throughput, and be used to support toll collection and road condition monitoring. At intersections, cameras could see cross-traffic and communicate this to oncoming vehicles to reduce accidents, as well as to extend crossing times for elderly or disabled people. The US DOT has announced plans to equip 20% of highways and 25% of traffic lights in the 75 largest metropolitan areas by 2028 with the capability to communicate with nearby vehicles that are also suitably equipped, and two automakers have already committed to supporting this. By 2036, the goal is for more than 75% of the intersections and traffic lights in major urban areas to have V2I capability [4.16].

V2P and V2C communications allow vehicles to detect the presence of pedestrians, cyclists, or other vulnerable road users even if they cannot be seen because of weather or obstacles, such as when a pedestrian walks across the street from behind a parked van. It can be used to alert the vulnerable road user's smartphone or wearable (perhaps embedded in a bicycle helmet), and vibrate to warn them to be careful. Because more than half the road fatalities worldwide are people "outside" the automobile, this could significantly improve safety and awareness for drivers, cyclists, and pedestrians, especially in urban environments or in areas and at times with limited visibility.

These V2X possibilities aim to enhance safety, efficiency, connectivity, and sustainability in the transportation ecosystem. While some applications are being implemented, others are still in the research and development phase. The successful deployment of V2X technologies requires standardization, protocols, cybersecurity measures, and a robust communications infrastructure, providing low latency and high coverage.

Cybersecurity concerns are also paramount and must be addressed to ensure the safety and integrity of V2V communication (both in terms of trusting the information sent, and in protecting vehicle systems from incoming messages). Blockchain can help with cybersecurity because it can ensure the integrity and authenticity of messages exchanged between vehicles, reducing the risk of data manipulation or unauthorized access. It can also enable the detection of malicious vehicles in V2V networks. By maintaining a distributed ledger of vehicle identities and behavior, blockchain can help identify and isolate vehicles that engage in malicious activities, enhancing the overall security of the network. A distributed ledger also eliminates the need for a centralized authority, reducing bottlenecks and improving the overall efficiency of the network.

Due to the fact that future vehicles will be connected to "everything," it is conceivable that they may have not just one or two cellular modems, as in connected vehicles, but maybe even three:

- Enterprise or business services covering OTA updates, remote diagnostics, and fleet management.
- Consumer-focused infotainment, navigation, personal device connectivity, and Wi-Fi hotspot.
- Safety-critical performance enabled by V2X to support ADAS and future autonomous driving.

This approach will improve security by separating safety-critical systems from consumer access and allowing optimized performance for each specific use case, as well as providing extra redundancy and reliability. However, it increases complexity, power consumption, and cost, and could be challenged by a software-defined approach that uses fewer physical modems but has robust virtualization to separate functions.

4.9.
eVTOLs (Air Taxis)

Maintaining, or even improving, traffic flow while adding AVs to the roads may be possible if the extra vehicles operate at times of the day when the roads are not heavily used. For delivery of goods that are not urgent or time-of-day dependent, it may be possible to operate during the night, if certain accommodations for receiving deliveries are permitted and made. This will be covered further in Chapter 7, Section 7.5.

During the day, however, a different approach is needed and an alternative emerging solution that may reduce congestion could be to shift some road traffic to the air. Several of the established automakers, such as Toyota, Mercedes-Benz, and Stellantis, are partnering with, or investing in, companies that are developing air taxis and eVTOL aircraft, and Hyundai even owns a subsidiary operating in this space [4.17]. This is not surprising because these vehicles could play a significant role in complementing ground transportation in the future in a cheaper, cleaner, and quieter manner than helicopters, making them better suited to operating in crowded cities. The eVTOL promises to combine the take-off capabilities of a helicopter with the efficient wing-based advantage of airplanes for travel. Since take-off can consume a quarter of a plane's fuel for flights under 500 miles there is a minimum distance, approximately 25 miles, beyond which an eVTOL may be more efficient than driving a car [4.18].

The James Bond movie *Thunderball* (1965) introduced "Little Nellie," a miniature personal helicopter. Unlike free video calls to anywhere in the world from a handheld device, this vision of the future has failed to materialize until now for a variety of reasons, including safety and cost. Sixty years later, we are beginning to have widespread deployment of drones, and eVTOL aircraft are beginning to emerge. There are, basically, two types of eVTOLs: those that have wings and those that only have rotors. The former may have an advantage for long-distance flights such as inter-city and airport-to-airport connections, whereas the latter may be best used inside the city for tourism, emergency services, businesses, and so on.

The main performance or technical challenges facing air taxi commercialization include limited payload and range, as well as a need to reduce noise levels and withstand air strikes and wind gusts. Despite having a limited 100-mile range, because batteries have only one-fortieth as much energy per unit mass as petroleum-based fuels, there are many applications where this range can still be useful. For example, some airlines are investing and partnering with eVTOL developers because they understand that the two can provide an integrated "door-to-door," not just "city-to-city," flight service; Delta Airlines' partnership with Joby Aviation is just one example. Advanced Air Mobility (AAM) solutions, such as air taxi transportation services and eVTOLs, are typically designed for short-haul transportation within urban areas and rely on vertiports to serve as takeoff and landing points. These might be located on the roofs of tall buildings to enable relatively quick transfers between ground and air transport. This is not without challenge, as rooftops might need reinforcement to support the additional eVTOL weight, and elevators may need to be added to reach the rooftop. AAM could ease congestion on the roads, deliver urgent freight, and support emergency services. From a commercial perspective, eVTOLs could be used for sightseeing and reduce travel times between the airport and downtown for wealthy businesses and consumers by more than half [4.19]. According to analysts at Morgan Stanley, the eVTOL market could reach $1.5 trillion annually by 2040 by catering to airlines, logistics, emergency services, agriculture, tourism, and security operations [4.20].

Unlike helicopters, eVTOLs are electric and could be relatively quiet and have zero emissions, but the broad adoption of air taxi services and eVTOL aircraft depends on them being certified by regulatory authorities. Certification requires not just type certification involving safety of the vehicle and the company's safety processes but also that the aircraft is designed to meet noise standards, compatible with landing facilities, harmonized with certification from other regulatory authorities, and capable of safe and secure operation. For the time being, the Federal Aviation Authority (FAA) in the US is requiring a pilot to perform a typical role, but the clear vision is for eVTOLs to become autonomous, which will create additional regulations (**Figure 4.18**).

Figure 4.18 In addition to goods movement, air taxi services may initially transport people from airports to city centers using eVTOLS.

As with robotaxis, China may be leading in eVTOL development and commercialization because the Chinese government believes it can create good jobs, drive innovation, grow exports, and stimulate the economy. It is, for example, supporting low altitude "demonstration zones" and establishing flightpaths around Shenzhen and Nanjing, that are complete with takeoff and landing pads. The Civil Aviation Administration of China (CAAC) is also relatively accommodating as US and European companies face more stringent regulations and safety protocols from the FAA and the European Union Aviation Safety Agency (EASA), and this has slowed progress when testing has been completed and operational deployment begins. Chinese companies also benefit from battery technology leadership and intense competition between EV makers, such as Xpeng, Geely, and SAIC, and within China's logistics sector to improve efficiency and meet the growing demand. In February 2024, AutoFlight, a leading flying taxi company, that has been developing a five-seater Prosperity eVTOL air taxi completed "the world's first inter-city electric air-taxi" flight between Shenzhen and Zhuhai in southern China [4.21]. The 50 km trip across the Pearl River Delta took only 20 minutes. Meanwhile, Chery demonstrated a flying car prototype in October 2024 that had flown 80 km on a test flight. It consists of three parts (aircraft, cockpit, and chassis), and it has been designed to take off and land vertically to avoid traffic congestion in dense urban areas (**Figure 4.19**).

Figure 4.19 EVTOLs, or air taxis on display.

References

4.1. Andrews, M., "CATL Announces Second-Generation Sodium Battery, Normal Discharge at −40°C," CarNewsChina.com, November 18, 2024.

4.2. Dunning, B., "What They're Saying about Electric Cars Now," Skeptoid Podcast #844, August 9, 2022, https://skeptoid.com/episodes/4844.

4.3. Ulrich, L., "Mercedes Proposes 'In-Drive Brakes' for Better EVs," SAE Automotive Engineering, March 2025, 21.

4.4. WBCSD Website, "Tire Industry Project," https://www.wbcsd.org/actions/tire-industry-project/.

4.5. J.D. Power Press Release, "Problems Plague BEVs despite Traditional OEMs Leveling Playing Field with Tesla, J.D. Power Finds," June 27, 2024, https://www.jdpower.com/business/press-releases/2024-us-initial-quality-study-iqs.

4.6. Rivers, S., "Huge Study Shows EVs More Reliable than ICE Cars with One Surprising Common Issue," Carscoop, April 13, 2025, https://www.carscoops.com/2025/04/new-study-shows-evs-are-more-reliable-than-ice-cars/.

4.7. Montoya, R., "What Is Bidirectional Charging and How Does It Work?," Edmunds.com, September 3, 2024, https://www.edmunds.com/electric-car/articles/bidirectional-charging.html.

4.8. Boehm, A. and Kraus, M., "The Rolling Chassis from Schaeffler Spearheads New Mobility Solutions," Schaeffler Group Colloquium 2022, https://www.schaeffler.com/en/media/dates-events/kolloquium/digital-conference-book-2022/rolling-chassis/.

4.9. Foglets.com, "Computation: Human Brain versus Supercomputer," April 10, 2019, https://foglets.com/supercomputer-vs-human-brain/.

4.10. Bobylev, D., "BYD Released the New 'God's Eye' Driving Assistance System," CarNewsChina.com, February 10, 2025, https://carnewschina.com/2025/02/10/byd-released-the-new-gods-eye-driving-assistance-system/.

4.11. Yole Group, "Yole Group Viewpoint – Automotive Semiconductor Innovation: Driving for the Next-Gen ADAS," March 20, 2024, https://www.yolegroup.com/yole-group-actuality/yole-group-viewpoint-automotive-semiconductor-innovation-driving-for-the-next-gen-adas/.

4.12. Wu, T. et al., "How China Will Help Fuel the Revolution in Autonomous Vehicles," January 25, 2019, https://www.mckinsey.com/industries/automotive-and-assembly/our-insights/how-china-will-help-fuel-the-revolution-in-autonomous-vehicles.

4.13. Burnett, S., "Hyundai Mobis: Holographic Dashboard Will Differentiate SDVs," Automotive World, March 27, 2025, https://www.automotiveworld.com/articles/hyundai-mobis-holographic-dashboard-will-differentiate-sdvs/?mc_cid=40b8f6fd01&mc_eid=c5e39e083a.

4.14. Burnett, S., "Continental: E Ink Technology Can Help Differentiate SDVs," Automotive World Website, April 8, 2025, https://www.automotiveworld.com/articles/continental-e-ink-technology-can-help-differentiate-sdvs/?mc_cid=1b6c2c8d4b&mc_eid=c5e39e083a.

4.15. Grant, A., "Under the Influence," ADAS & Autonomous Vehicle International, September 2024, https://adas.mydigitalpublication.com/publication/?m=71150&i=826548&p=26&ver=html5.

4.16. US Department of Transportation, "Saving Lives with Connectivity: A Plan to Accelerate V2X Deployment," August 2024, https://www.its.dot.gov/research_areas/emerging_tech/pdf/Accelerate_V2X_Deployment_final.pdf.

4.17. Kroll, "Automotive Industry Report Spring 2024," https://media-cdn.kroll.com/jssmedia/kroll-images/pdfs/executive-summary-automotive-industry-insights-spring-2024.pdf.

4.18. Keoleian, G., "Role of Flying Cars in Sustainable Mobility," *Nature Communications* 10 (2019): 1555, doi:https://doi.org/10.1038/s41467-019-09426-0.

4.19. "eVTOLs - Roadmap to 2030 and Beyond," FutureBridge Report, August 2024.

4.20. Morgan Stanley Research, "Are Flying Cars for Takeoff?," Blue Paper, January 23, 2019, https://www.morganstanley.com/ideas/autonomous-aircraft.

4.21. AutoFlight Press Release, "AutoFlight Showcases World's First Inter-City Air-Taxi Flight," February 27, 2024, https://www.autoflight.com/en/news/shenzhen-zhuhai-route/.

Chapter | 05

Some Consequences of the Future Automotive Trajectory

5.1. Unintended Consequences

"Any solution at scale creates a new set of problems," illustrates the fact that while we may introduce technologies to solve certain issues, their widespread adoption and implementation can lead to a new set of challenges and some potentially unintended consequences. To consider the possible side effects of the disruptive automobile evolution, it is instructive to look at the widespread adoption of smartphones, because this is also a recent consumer product that has cutting-edge technology.

It is easy to list the benefits that have been delivered by smartphones. Instant connectivity with others and with a vast source of knowledge (Internet) anywhere around the world—for free—was science fiction a few decades ago. Star Trek introduced the idea of a single handheld device, and nearly all of us now have something similar that can provide navigation, mobile banking, health monitoring, gaming, music, photography, shopping, language translation, news, social networks, and much more!

However, in tandem with Wi-Fi and cellular communications technologies and smartphone apps, the following are some of the issues that have clearly emerged since the smartphone's introduction:

1. "Addiction," because smartphones provide stimulation and connectivity that can lead to compulsive behavior for some users and can interfere with study and work, as well as disconnecting users from their real-world social life.
2. Distraction, because apps, games, and notifications can reduce attention span and affect the ability to concentrate on tasks for extended periods.
3. Sleep disruption, because the light emitted from smartphone screens can disrupt natural sleep cycles, and night-time use can cause poor sleep.
4. Cyber-bullying, especially of children, because smartphones make it much easier to spread harmful content, often anonymously.
5. Privacy is lost when smartphones generate large amounts of personal data that can potentially be hacked, shared, or used for surveillance without the user's consent.
6. Spreading of misinformation, hoaxes, and harmful conspiracy theories is made far easier with smartphone usage and messaging.
7. Texting while driving is a particular automotive challenge because it can lead to accidents and fatalities.

It can be argued that many of these issues are not caused by the smartphone *per se* but by the combination of connectivity and content (apps). However, without the portable convenience of a handheld, these issues would likely not be so pervasive. In addition, there may yet emerge long-term health effects associated with continued smartphone usage, such as poor posture and social isolation because some children grow up feeling more comfortable looking at a screen than striking up a conversation with peers. Clearly, while enormously powerful and convenient, the addictive and ever-present nature of smartphones, when combined with the Internet and widespread connectivity, has created cultural and societal side effects that were largely unforeseen during their meteoric rise in popularity. Nowadays, the automobile is often referred to as a "smartphone on wheels," and automakers are trying to copy the profitable software, service, and data business model pioneered by tech companies and move away from just supplying a hardware product. With the sobering example of the smartphone in mind, this chapter will focus on future vehicles and see what "unintended consequences" or side effects may lie in store.

If we could go back in time to the early 1900s and walk around major town centers, we would probably be struck by the smell and sight of streets covered with horse manure, and this would not be pleasant, especially when walking in the rain and slipping! Horse-drawn carriages were the most common form of transport at that time, and one of the perceived benefits of the car, whether energized by batteries, gasoline, or steam engines, was that it was supposedly "clean" and did not pollute the streets. Within a generation the dominant form of transport on most city streets shifted from horse-drawn carriages to motor vehicles, as shown in **Figure 5.1**.

Figure 5.1 New York, 5th Avenue: 1900 and 1913.

National Archives and Records Administration, Records of the Bureau of Public Roads.

A few cars may seem clean, but, at scale, we now know they create significant air pollution because the combustion of fuel produces nitrogen oxides, particulates, and carbon monoxide, which can combine to form smog under certain atmospheric conditions. Moreover, the same combustion is a significant source of carbon dioxide, which we now know is a greenhouse gas. The production and disposal of vehicles, as well as the construction and maintenance of the roads needed for them, have also had a significant environmental impact, including the consumption of natural resources, habitat fragmentation, and the generation of waste.

There have been many other outcomes that were not obvious a century ago. For example, car usage enabled suburban sprawl, longer commutes, and the loss of land taken over for parking and roads. The dominance of gasoline-fueled vehicles has created a global dependence on oil, leading to geopolitical conflicts, price fluctuations, and economic instability. CAFE regulations to improve fuel economy may have unintentionally led to the emergence of larger vehicles, as was discussed in Chapter 2, Section 2.1.3. The increasing number of vehicles on roads, especially in urban areas, has led to severe traffic congestion, which causes delays, and increases stress levels for commuters as well as fuel consumption. With nearly every vehicle now also having a compelling distracting device (smartphone) inside it when moving, there is an increased risk of accidents, and while automobiles have many safety features, the sheer number of vehicles on the road and the potential for human error are still causing more than one million deaths every year all across the world.

Or consider the unintended environmental consequences that may arise from online goods delivery. While online shopping and delivery services can potentially reduce the number of individual trips made by consumers to stores, in many cases the opposite happens. For example, extra trips may be made by consumers for browsing in physical stores before ultimately making online purchases. A significant fraction of online purchases are also returned because it is harder to choose the correct product online and online retailers have made it easy to return stuff. If packages cannot be efficiently grouped together for delivery because the consumer wants to receive them quickly, then the increased number of delivery trips could offset the intended emissions reduction. Moreover, online platforms consume energy in their cloud operations, and there is much more packaging waste that must be taken into account.

We should expect that any solution, <u>at scale</u>, will create a whole new set of problems!

5.2. Systems Thinking

The problem is not necessarily the technology itself but how it is used and the broader implications of its adoption. It is important, albeit impossible, to anticipate all the potential consequences, especially negative ones, when evaluating the impacts of technological solutions at scale so that we can try to develop them in a more responsible manner, preparing the appropriate groundwork for their adoption, minimizing adverse outcomes, and maximizing the intended benefits for society and the environment. After all, we hope that such thinking is "happening" now with AI.

Emerging automotive technologies like autonomy, connectivity, and electrification are often portrayed through a narrow techno-economic lens as clean, efficient solutions to modern transportation challenges. However, this framing fails to account for the reality that transportation is a complex system that has been shaped by a mix of technical, social, political, economic, and environmental factors over many decades. Therefore, any technological disruption can and will create new vulnerabilities and tensions in this system. For example, EVs can eliminate tailpipe emissions and diversify energy sources away from petroleum dependency, but their net impact will depend on how the electricity is generated and how the battery materials are mined. Autonomous mobility could increase accessibility, facilitate links with public transport, and generate new residential apartments from unused parking lots, but it can also induce new travel demand, urban sprawl, traffic congestion, and energy usage.

Rather than implementing transformative technologies because "we can," a more prudent approach asks "why," "for whom," and "in what way" their adoption will potentially affect materials and energy resource flows, human capabilities, and politics across society.

Technologies should be carefully integrated not as singular solutions, but as components of a holistic mobility system that are tailored to the affected population's needs and values. Let us consider several automotive technologies and how the expected benefits could, without thoughtful adoption, be outweighed by a minefield of concerns.

5.3.
EVs

In the past, the main discussion about EVs' harmful effect on the environment was centered on how the electricity to power them was generated. EV skeptics argued that if coal was used to generate electricity for EVs, then the overall, or "well-to-wheels," pollution and greenhouse emissions could actually be worse than with ICEVs, conveniently overlooking the health benefits of transferring an equivalent level of pollution from a heavily populated area to a remote powerplant. The fact that the electricity mix continues to become cleaner each year, as renewable energy becomes cheaper and as coal-fired plants are shut down, has made this far less of a potent argument. However, the reality of battery manufacturing at scale has recently emerged as an argument against EVs, despite oil drilling and oil spills creating significant environmental issues upstream of the vehicle's production. Argonne National Lab's Greenhouse Gases, Regulated Emissions and Energy Use in Technologies (GREET) model has calculated it takes approximately 13,500 miles before an EV breaks even with an ICEV in terms of net greenhouse gas emissions contribution [5.1]. This is only around one year of driving for the average motorist. This mileage clearly depends on the battery chemistry and source of electricity used for vehicle recharging, but, over time, both should become significantly cleaner with lighter batteries, greater adoption of battery recycling, and with renewable energy producing the electricity.

5.3.1.
Environmental, Social, and Governance (ESG) Issues

Figure 5.2 ESG concerns with battery materials production: water depletion with lithium mining (Chile), water pollution with nickel mining (Indonesia) [5.2], and child labor with cobalt mining (DRC) [5.3].

In addition to environmental concerns with battery mining (**Figure 5.2**), there are also human rights abuses, child labor, and exploitative mining practices in many parts of the world where minerals are located. Examples, according to Infyos, include cobalt mining in the Democratic Republic of Congo where five-year-old children mine in hazardous conditions, and people in the Xinjiang Uyghur Autonomous Region (XUAR) in northwest China who may be forced to work in lithium refining facilities. Infyos estimates that three-quarters of the global EV battery materials supply chain may be "contaminated" by these labor law violations [5.4]. Responsible sourcing for battery materials will require robust auditing, "fair trade" certification, and mineral traceability as well as more direct investment in environmentally and socially responsible mining operations. Many automakers have stated their commitment to human rights due diligence, but according to Amnesty International, they do not provide much evidence to support these claims in their supply chain [5.5]. We all want battery costs to come down, but it should not be achieved by exploiting labor or the environment.

5.3.2.
National Security Issues

China currently dominates the global lithium-ion battery supply chain to an unprecedented extent, controlling a large share of the raw materials refining, the processing into cell components (e.g., anode, cathode, electrolyte, etc.), and the manufacturing of complete battery cells for use in EVs. To put this into numbers, China accounts for nearly 90% of global installed cathode active material manufacturing capacity, more than 97% of anode active material manufacturing capacity, and roughly 80% of global raw material refining capacity for battery metals, and in 2022, China processed approximately 65% of the world's lithium, 74% of cobalt, and 100% of natural graphite for batteries. Nearly 90% of the world's processed rare-earth permanent magnets, widely used in electric motors for propelling EVs, come from China.

These dependencies on China, illustrated in **Figure 5.3**, raise several concerns for other countries [5.6]. For example, any political or economic instability in China could disrupt the supply chain, impacting EV production globally. Even if China is stable, there is a risk that they can manipulate battery prices, through policy changes or export restrictions, impacting the cost of EVs for consumers worldwide and creating an uneven playing field for automakers in different regions. And, of course, tensions between China and other countries could lead to targeted restrictions on battery exports, further disrupting the supply chain and hampering other countries in their efforts to grow a domestic EV ecosystem and transition toward cleaner transportation. Experience with the Organization of the Petroleum Exporting Countries (OPEC), where there is far less concentration of the energy supply chain, shows how vulnerable the world's economy can be to manipulation, price setting, and supply reliability.

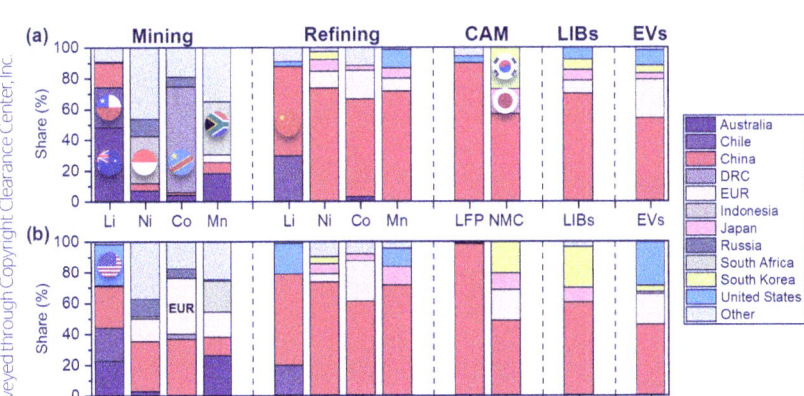

Figure 5.3 China's dominance of the battery materials supply chain. a) Geographical distribution, b) ownership distribution.

To diversify the supply chain and become less reliant on China, it will be necessary for governments to help develop domestic extraction, processing, and refining capabilities for lithium and other critical minerals while also addressing environmental concerns. Secondly, international partnerships, in the form of joint ventures and long-term supply agreements, may also need to be forged between friendly countries that produce the materials needed for making batteries. Strategic mineral reserves will need to be established for critical battery materials that could provide a buffer against supply disruptions and price volatility. These reserves could be maintained through domestic stockpiling or through international agreements and shared stockpiles with partner nations.

At the corporate level, supply chain risks can be reduced through partnerships between EV manufacturers and mining companies, perhaps even leading to vertical integration for EV makers across different stages of the supply chain (raw material extraction, materials processing, battery and EV manufacturing). This could be achieved either through acquisition, investment, or partnership with raw materials suppliers to secure a stable supply of critical materials. For example, GM has invested $650M in a US-based lithium mine in Nevada in exchange for exclusive rights to the output. In addition to increased supply chain security, this approach can also lower costs by eliminating intermediaries. However, it is not risk-free because of commodity price fluctuations and the need to manage environmental and geopolitical impacts.

Extensive battery recycling can, of course, help with all of the environmental and security challenges mentioned above. Investing in advanced battery recycling technologies and establishing efficient recycling systems should help to recover valuable materials from spent batteries, reducing the demand for new raw materials by up to 40% mid-century, according to the International Energy Agency [5.7]. For example, Glencore has proposed Europe's largest EV battery recycling plant in Sardinia and will recover lithium, nickel, and cobalt for reintegration into the battery production cycle [5.8]. This not only addresses environmental concerns with battery disposal but will also create an attractive market for recycled materials that can mitigate the need for mining new materials.

Promoting the reuse and second-life applications of EV batteries could further extend their life cycle and contribute to a more circular economy. However, some developments are going in the opposite direction. Recently, several automakers have begun to integrate battery cells either directly into packs (cell-to-pack or CTP) or, going one step further, even directly into the vehicle body (with Tesla, BYD, and Leapmotor being leaders) instead of the usual approach of integrating battery cells into modules into packs into the vehicle. This tighter integration makes recycling more difficult, but the CTP approach eliminates wasted space with module and pack housing units, allowing more batteries to be packaged into a given space. It works better for LFP batteries because they have a more stable structure than NMC, allowing them to be more tightly packed together in a CTP design with less concern for overheating, whereas NMC cells need more space between them for safety reasons. When combined with CTP, the LFP chemistry's energy density can almost approach that of NMC in the final battery pack but with a lower cost. CTP also reduces vehicle mass, but the tighter packaging can make servicing, repair, and recycling more difficult, and if the automaker is not vertically integrated and relies on battery suppliers, then it becomes harder still. As battery cost and performance continue to improve, such "desperate" measures to extend range may become unnecessary, and it should be possible to integrate future batteries into the vehicle so that they can be easily refurbished, or reused in other applications (such as ganged together for grid storage or broken down into small e-kits for the type of low-speed, short-range vehicles discussed in Chapter 8, Section 8.6.2) or recycled into their constituent materials at the end of life. This will be easier if the automakers or OEMs form joint ventures with battery suppliers because this allows the automaker to exert more influence on the battery design in the early stages of development, and to design the battery with recycling in mind.

Another development that may occur, driven by battery recycling and cybersecurity protections, is that automakers may move toward leasing vehicles, instead of selling them, to consumers. GM did this with the EV1 during the 1990s for a variety of reasons. It allowed GM to recall the vehicles more easily, limited their potential liability if issues arose, and helped manage uncertainty around battery life. If, as is currently being proposed in California's bill SB 615, EV batteries must be recovered at the end of their useful life, then automakers may need to ensure access to and control over the vehicle at all times, making leasing their preferred option. Moreover, as the threat of sophisticated cyberattacks over the life of the vehicle increases, it is possible that the vehicle's original chip hardware and software protections will need to be upgraded, and hardware repair or replacement is easier to achieve if the automaker, not the consumer, owns the vehicle.

5.3.3.
EV Mass Issues

Today's crossover-based EVs typically weigh approximately 10% more than the ICEV equivalent, and the larger batteries used in electric pickup trucks can make EVs weigh up to 1000 kg more than a comparable-range gasoline pickup truck. With the current state of battery technology and a consumer-auto industry mindset that dictates 300 miles of range capability, adverse effects caused by vehicle mass are going to be compounded by the shift toward EVs.

For example, heavier vehicles pose more significant risks to other cars, cyclists, and pedestrians in the event of a collision because heavier vehicles have more momentum, which can lead to longer stopping distances and more severe injuries. Heavier vehicles also tend to have a stiffer construction so that they absorb less energy in a collision. In a collision with a pedestrian, for example, more force is transmitted to the pedestrian and this leads to more severe injuries. Third, heavier vehicles, like SUVs and trucks, tend to be taller and have a flat hood line instead of one that slopes down. This means that they hit pedestrians, cyclists or cars at a point higher up on their bodies, potentially causing more severe internal injuries. A car's sloping hood can allow some "roll" of the pedestrian onto the hood, which extends the deceleration time but a tall SUV's front end will tend to cause a more abrupt stop. Taller front ends also create larger blind spots immediately in front of the vehicle, increasing the risk of collisions occurring in the first place. In principle, using a skateboard platform and eliminating the "engine compartment" creates a front end that can be lower and less rigid. Most automakers, however, are retaining the style and adding storage space, or a "frunk" instead.

A less noticed EV mass issue is that additional weight increases wear on tires and the amount of tire particles released into the atmosphere (which can end up as microplastics in the oceans). Extra weight will also increase the load on roads, bridges, and parking structures compared to conventional ICEVs of a similar size. Many roads, bridges, and parking structures were designed and built decades ago when vehicles were generally lighter, so that the increased weight from heavier EVs can accelerate wear and tear on this aging infrastructure, potentially leading to cracks, potholes, and structural degradation over time. In extreme cases, if the infrastructure is already compromised or weakened, the additional weight from heavy EVs could potentially contribute to catastrophic structural failures, and even if this does not occur, the increased stress from heavier EVs will likely require more frequent and costly maintenance and repairs for roads, bridges, and parking structures. Residential streets, for example, are designed to last 20 years but may need resurfacing every ten years if vehicles are heavier, and increased goods delivery, using autonomous electric vans and trucks in the future, will exacerbate this further. Street residents, municipalities, and taxpayers will end up paying for this extra cost, which can typically range from $0.5M/mile for a side street up to $5M/mile for one lane of a highway. To make matters worse, a significant portion of funding for road maintenance and construction comes from fuel taxes collected on gasoline and diesel sales. As more drivers transition to EVs that currently do not pay fuel taxes, and as ICEVs become more efficient, the revenue stream for infrastructure maintenance could drop so much that there will need to be some new revenue policies enacted, such as implementing weight-based road usage fees or taxes for EVs.

However, from an infrastructure perspective, it is not all bad because EVs can help stimulate renewable energy by consuming energy generated by the sun and wind, notably during the daytime. EVs are now being developed with bidirectional charging capability (Chapter 4, Section 4.2), meaning they can also send power back to the grid. This can stabilize the grid by acting as a huge energy storage buffer whereby EVs can provide additional electricity when needed by the local utility—if everything goes according to plan. However, more people are likely to raise the issue that large numbers of EVs will put more strain on the grid, increasing the number of brownouts with a new risk that power outages will make EVs inoperable.

While EVs represent a critical technology for reducing transportation emissions, their reliance on certain mineral resources raises environmental, social, and geopolitical concerns that require holistic stewardship across the complete battery life cycle. The weight-related issues mentioned above may disappear over time as newer batteries pack more energy and range into a given mass, but what is abundantly clear is that smaller batteries help reduce all of these concerns (environmental, national security, stress on the infrastructure) as well as making vehicles more affordable and efficient. The idea of right-sizing the battery and the vehicle forms an essential part of the new framework proposed in this book, and will be discussed in Chapter 8, Sections 8.4.1 and 8.4.2.

5.4. Connected and SDVs

As covered in Chapter 3, Section 3.4 the first connected vehicles were introduced in the 1990s by GM and co-branded as Onstar. The original focus and value proposition of connectivity was to provide emergency calling in the event of an accident. Other connected vehicle features, such as turn-by-turn navigation and vehicle diagnostics, were added much later. Nowadays, more than half of all new vehicles sold each year have cellular connectivity and Wi-Fi capability, and it is expected that nearly all new vehicles sold in 2030 will have this level of connectivity. Connectivity is well understood to be a "double-edged sword," and several issues are created with connected vehicles that can overwhelm the benefits for those unlucky enough to be on the receiving end.

5.4.1.
Privacy

In 2023, the Mozilla Foundation published a report on privacy practices in the car industry [5.9]. It found that 25 car brands were collecting more data than was necessary, that 84% were sharing or selling driver data, and that 68% of the brands had been hacked or had leaked data since 2020. Volkswagen, for example, has exposed information from 800,000 EV owners because of a flaw in its software.

GM's Smart Driver program was rolled out in 2023 and was designed to monitor and potentially influence driver behavior by collecting data on acceleration, braking, cornering, and speeding events. The program aimed to offer discounts on insurance premiums based on the collected data, with the intention of promoting safer driving habits. Perhaps the expectation was that safe drivers would opt in, and drivers who knew they might not get an insurance premium discount would either not enroll or would opt out. However, in practice all users were enrolled in the program without even being asked to opt-in, meaning the automaker was collecting data without their approval, monitoring their every move and how they drove, and selling this information, via a broker, to insurance companies that might then raise premiums. Drivers are understandably worried about how these data might be used by automakers, insurance companies, and even law enforcement. Because of the ensuing public outcry, GM now requires users to opt-in to data collection programs and as part of a settlement agreement with the Federal Trade Commission (FTC), GM cannot share sensitive data to consumer-reporting agencies for five years and must be more transparent to consumers over the collection, use, and disclosure of their connected vehicle data.

As mentioned in Chapter 4, Section 4.6 Waymo is capturing data from the interior occupant-monitoring camera in its robotaxis to train customer-facing AI models for safety, obeying rules, cleanliness, and help in case of emergency. It may also use its face recognition software to determine what ads to sell in order to generate extra revenue from Waymo riders. This should, of course, be an opt-in process but monetization of the surveillance economy will probably mean it is opt-out instead, a fact of life we have become accustomed to with smartphone apps, and which will slowly but surely carry over into the vehicle.

Welcome to the connected vehicle "dark side." As consumers, we are highly vulnerable and, in return, we receive access to cool features that make life a little more convenient when all goes well! Connected vehicles rely heavily on data collection and communication with cloud services or the manufacturer's servers. This raises several concerns about the privacy of personal data, such as location tracking, driving habits, and potentially sensitive information (**Figure 5.4**). The location data that tracks movement could be used by companies to advertise their products or services, generating a new revenue stream for the automaker in the process. Automakers say that personal data are not shared because they are anonymized. This may not be completely reassuring because location data can usually be reidentified with individuals as we tend to have unique driving patterns. Information can also be inferred about home and work address, a child's school location, daily driving routines and planned destinations, as well as religious, rehabilitation, and medical facilities, to cite a few examples. Moreover, tracking is also enabled for the vehicle owner, and there have been multiple cases of "stalking" by a spouse during divorce proceedings, with Mercedes-Benz even refusing to disable the feature when requested by the police. Tracking could even be abused by law enforcement without proper regulation.

Figure 5.4 The "dark" side of connected vehicles.

As occupant monitoring systems become ever more sophisticated and pervasive (and invasive), it will be necessary to ensure that the collection and analysis of personal data are handled responsibly and securely so that privacy concerns are not violated. The stakes become higher when health data are involved, which may be more common in the future, as was discussed in Chapter 4, Section 4.7. Health monitoring inside the vehicle can be beneficial because it can be used to track health daily in a somewhat controlled and similar environment each time, making it potentially useful for ongoing measurement of chronic issues. However, these health data should only be shared with the person's explicit permission, and the accuracy and reliability of health monitoring systems need to be carefully evaluated because false positives or negatives could lead to unnecessary alerts or missed detection of health issues, potentially impacting occupant trust and safety.

5.4.2.
Security

There is also a risk of data breaches or unauthorized access to this sensitive information by hackers or other malicious actors who can sell this information or use it for identity theft. A successful hack on a connected vehicle could lead to the hacker being able to spoof GPS signals or take control of the vehicle's braking and steering system and put the driver and passengers at risk of a serious accident. Hackers could also lock drivers out of their vehicles and demand a ransom to regain control. Cloud-based storage and sharing of vehicle and owner data introduce additional risks of cybercrime or accidental data leaks that can damage the automaker's reputation and business (e.g., lost customers and paying fines for privacy violations).

An SDV may require a constant Internet connection to receive software updates, access certain features, and communicate with the manufacturer's servers so that any loss of connectivity or service disruptions could disable certain features or functionalities, affecting the driving experience. A software bug or glitch, that might even be introduced with remote software updates, could have the same effect. Greater use of digital assistants, such as GM's Onstar Virtual Assistant, can provide customer benefits, but they have also been known to enable unauthorized commands, which can potentially allow the vehicle to be stolen and offer unauthorized remote control and data theft, posing safety risks and privacy concerns.

At the international and government level, a large-scale cyberattack on connected vehicles could have a crippling effect on the transportation infrastructure. Hackers could potentially disrupt traffic flow and the flow of essential goods and services, causing chaos and threatening public safety. Even the electric infrastructure is at risk, as EVs become more common. Moreover, if not properly secured, connected vehicles could be used by foreign actors to steal sensitive or classified data concerning government officials, military movements, or critical infrastructure. These concerns are leading to the likelihood that the US will ban Chinese connected vehicles or impose restrictions on automakers using Chinese-supplied connectivity-related technology, for fear that the software and hardware inside future imported vehicles from China could track Americans' locations and report back to Beijing.

In response to all of these concerns, companies are implementing robust cybersecurity measures such as regular software updates, encryption, and intrusion detection systems. To address cybersecurity concerns, manufacturers and regulatory bodies need to prioritize robust measures, data privacy protections, and consumer rights regarding ownership and control over their vehicles. Clear guidelines and standards should be established to ensure transparency, safety, and accountability in the development and deployment of SDVs. Given its importance, we can expect that cybersecurity will become a highly regulated standard feature, like airbags. An automaker's cybersecurity reputation is likely to be judged by consumers and could be a major brand purchase criterion.

5.4.3.
Life Cycle Concerns

In principle, SDVs could enable the mass production of large numbers of vehicles all having the same hardware. It would allow automakers to sell software updates to wealthy consumers for unlocking premium features and capabilities, while allowing less affluent consumers to have the same basic vehicle at a lower cost. This approach could lead, in effect, to wealthy customers subsidizing other vehicle buyers. However, this would also burden all vehicles with the same hardware, which would add mass and could cause more costly repairs. For example, a fender bender becomes more expensive to repair if it contains sensors that need to be replaced and recalibrated, even if those sensors have not been software-enabled on the basic vehicle.

For this reason, SDVs are likely to cost more initially because they contain the advanced hardware components required for future software updates. SDVs will likely rely heavily on software OTA updates for new features and bug fixes that can be delivered without requiring hardware changes. Subscription fatigue may occur, whereby features that may now be standard (like AEB) will become subscription-based in the future. This could add significant ongoing costs to car ownership, but recurring revenue is very attractive to automakers if they can achieve it.

As the software becomes increasingly personalized, it may also become more difficult to switch to another manufacturer, as we know already happens with smartphones where consumers are "locked in" to a company because the switching costs become too high. Like smartphones and other software-based devices, SDVs may become obsolete more quickly because of the rapid pace of software and technology changes, meaning that automakers may discontinue support or updates for older models, rendering certain features unusable or the vehicle less functional over time. The vehicle should continue to drive, but some of the original vehicle's computer hardware may need to be replaced, perhaps by a dealer, because it is not powerful enough to run the new software, as has been happening with Tesla's FSD rollout. There is also a risk of planned obsolescence, where manufacturers intentionally limit the lifespan of vehicles to encourage more frequent upgrades or purchases. Safety regulations and industry standards might need to be established to ensure that vehicles remain functional for a reasonable period without mandatory hardware upgrades. In contrast to smartphones, one hopes that the core functionality of the vehicle (starting, driving) will still be operational so that the customer is not forced to buy a new vehicle!

And, finally, there is the risk that the automaker files for bankruptcy, as happened in October 2023 in the US with Fisker Inc. It also happened in China with WM Motor [5.10], a Shanghai-based EV maker, because there are far more car companies in China than the market can support. The tech industry tends to be a "winner-takes-all" business with very few players in each sector, and this may happen more and more in the auto industry, accelerating consolidation and obsolescence of certain brands. When this happens, software support disappears and the smartphone app no longer works, meaning that some features like remote door unlock and cabin precooling may no longer be available, and other features that were meant to be provided as a software update after vehicle sale did not materialize. This raises the concern about long-term software support, particularly as cars can often last more than ten years. This risk could make it difficult for unproven startups to sell new vehicles and even legacy automakers could go out of business in a decade, given the disruption to the auto business. In short, the issue of guaranteeing software support will need regulation and may be one more reason for automakers to move to a two- or three-year lease model (in addition to ensuring access to battery recycling and upgrading cybersecurity protections).

5.5. ADAS

Human error is responsible for, perhaps, over 90% of vehicle collisions, according to a frequently cited NHTSA 2015 paper [5.11-5.13], as shown in **Figure 5.5**.

Figure 5.5 Causes of vehicle-related crashes [5.11], based on a 2005–2007 survey conducted by NHTSA. The most common driver errors are caused by lack of recognition, often caused by distraction, and poor decision-making, such as driving too fast.

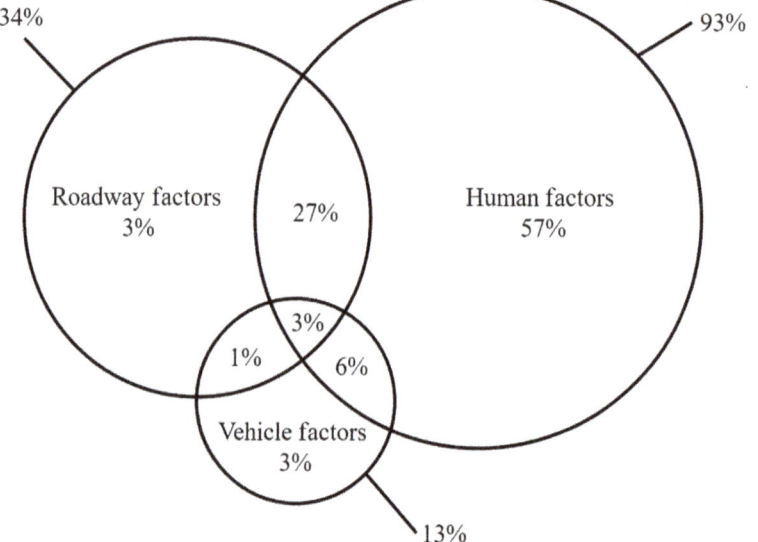

Critical reason attributed to	Estimated	
	Number	Percentage* ± 95% conf. limits
Drivers	2,046,000	94% ±2.2%
Vehicles	44,000	2% ±0.7%
Environment	52,000	2% ±1.3%
Unknown critical reasons	47,000	2% ±1.4%
Total	2,189,000	100%

In principle, ADAS can reduce road accidents and fatalities. For example, a DMS could reduce recognition errors by alerting the driver when they are distracted or drowsy. ADAS can also be used to warn the driver to remain at a safe distance behind another vehicle or not to exceed the speed limit, as with intelligent speed assist, which has recently been mandated in Europe. As ADAS moves from alerting the driver to driving the vehicle autonomously under certain conditions (partial or conditional autonomy), the vehicle can potentially manage the driving more safely than when relying on the driver. AEB systems are a good example of this.

However, while ADAS technologies are designed to enhance safety by assisting drivers and reducing the risk of accidents, there is a legitimate concern that they may encourage less attentive driving behavior if the driver trusts them too much to the point where they pay less attention to driving. In a similar vein, connectivity enables access to various services and information while driving and can improve safety, but it can also contribute to driver distraction if not used responsibly. In short, what we have today is an interim step of limited autonomous driving and the ambiguity is creating several new safety issues. ADAS technology may not be delivering the safety benefits expected by the suppliers developing them, the automakers that are integrating them into new vehicles and the government regulators who are expecting them to reduce road collisions and fatalities. Let us consider the cost:benefit equation one at a time.

5.5.1.
Cost

ADAS promises a future of safer roads, but the current landscape surrounding ADAS human-machine design, associated user challenges, and cost implications paints a more nuanced picture. In terms of a cost–benefit analysis, ADAS typically adds the following to the price of a vehicle:

- Basic ADAS (e.g., AEB, LDW): $200–$500
- Mid-level ADAS (e.g., ACC, blind spot monitoring): $500–$1500
- Advanced ADAS (e.g., highway assist systems, automated parking): $1500–$3000+

The price escalation is due to the extra sensor content needed as well as consumer willingness to pay for convenience rather than safety. Moreover, the costs associated with ADAS should also consider repair. Modern automobile windshields are no longer simple sheets of laminated glass but contain a myriad of other vehicle systems, such as rain sensors, heating elements, antenna lines, and optical coatings to reduce infrared transmission. Increasingly, the windshield also has cameras and other sensors mounted to it for ADAS features like lane keep assist and AEB. A stone chip thrown up by a passing truck may cause this "expensive" windshield to need to be replaced. The repair cost is even higher than expected because the sensors must be properly calibrated after a windshield replacement, and this can require several hours of labor at a rate over $100/hour. Replacing a bumper or side mirror can also be expensive for a similar reason that they contain ADAS components, such as radars, ultrasonics, and cameras, which must be calibrated after replacing or repairing.

This means that even if the ADAS sensors reduce the frequency and severity of collisions, the premiums may still go up because insurance companies factor in the higher cost of these repairs. A 2023 study by American Automobile Association (AAA) found that ADAS features can increase repair costs by up to 37.6% after a crash because of replacement and calibration needs [5.14]. However, the Insurance Information Institute (III) reports that auto insurance discounts can range from 5 to 10% for vehicles equipped with ADAS, but the long-term impact of ADAS on insurance rates is yet to be determined. As the technology matures, accident rates might decrease, potentially leading to lower insurance costs in the future. Some insurance companies do offer discounts or lower premiums for vehicles equipped with certain ADAS. Examples of insurers include Progressive (for vehicles with forward collision warning [FCW], and AEB systems), Travelers (LDW), and USAA (for Blindspot Monitoring [BSM], and rear cross-traffic alert features). On the other hand, some insurance companies have increased the rates for vehicles with ADAS because of the higher repair costs associated with calibrating and replacing the sensors and components after a claim. Examples include AAA (vehicles with advanced safety packages like Tesla's Autopilot), and some insurers add surcharges or eliminate discounts for luxury vehicles with extensive ADAS sensor suites that could make vehicle repair so expensive that the vehicle is "totaled." Moreover, with AVs having varying degrees of human intervention, there may be additional costs and delays as insurers investigate if the vehicle or driver is liable for any collision.

5.5.2.
Effectiveness

The Insurance Institute for Highway Safety (IIHS) reported that FCW with AEB can reduce rear-end crashes by up to 50% [5.15], and a Swedish study estimated that LDW can reduce fatal car occupant head-on and single-vehicle crashes by 27–43% [5.16], but according to the United States Department of Transportation (USDOT) statistics, shown in **Figure 5.6**, traffic fatalities per 100 million VMT actually increased from 1.10 to 1.26 between 2013 and 2023 [5.17] even though ADAS penetration has increased significantly over the same time frame. This may be attributable to a variety of factors including vehicles becoming heavier and drivers (and pedestrians) becoming more distracted.

Figure 5.6 Traffic fatalities in the US per 100 million VMT (2013–2023) [5.16].

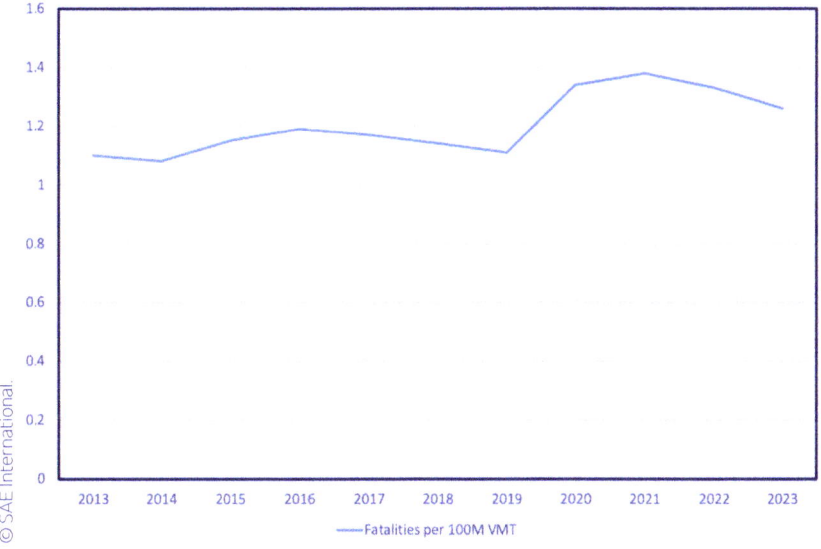

In fact, a 2021 AAA study found that some drivers using ADAS features exhibited riskier behaviors like tailgating or taking their hands off the wheel, potentially because of overreliance on the system [5.18], while another study by IIHS found that drivers are more likely to be distracted by L2 partial autonomy systems while still being able to "cheat" the DMS [5.19]. ADAS features vary significantly between manufacturers and models in terms of naming, capabilities, activation/deactivation methods, and user interfaces, and this inconsistency can make it difficult for drivers to learn and use them properly, increasing the risk of misuse.

Longer term, there are critical human factors that must be addressed to ensure safe and usable operation by humans in the automobile environment. Increased reliance on automation and connected features could lead to a decrease in driver skills and situational awareness. If, for instance, drivers become overly reliant on ADAS, they might not pay close enough attention to the road, potentially leading to accidents if the system fails. This might mean there needs to be on-road coaching to mitigate the degradation of driver skills. Recent research at the University of Nottingham studied 17 motorists (admittedly, a small sample) in a simulator and found that driving skills were inferior during the first 10 seconds after taking back control from autonomous operation, as evidenced by erratic maneuvers [5.20].

To help make ADAS easier to understand and use, it would help if the ADAS alerts were standardized across all vehicle types, brands, and models. On the other hand, some level of personalization and learning for each user may be desirable because, for example, different drivers prefer different following distances when using ACC. Unfortunately, this adds complexity, meaning that finding the balance between being easy to use and being useful is tricky. Perhaps an easier standardization that the automakers could achieve is to market the same feature with the same name as each other. For example, blind spot warning is called "Audi Side Assist" on Audi vehicles, "Blind Spot Monitor" on Toyota vehicles, and "Side Blind Zone Alert" on some GM vehicles [5.21].

It is ironic that a technology like ADAS, designed to enhance safety and convenience, is now requiring the introduction of another layer of technology, DMS, to detect if the driver is attentive, as illustrated in **Figure 5.7**. There is an inherent tension between the convenience offered by ADAS and the need to maintain driver engagement for safety reasons, and one can argue that DMS is effectively being designed to facilitate the convenience of ADAS in **not** driving while trying to achieve acceptable levels of safety. This highlights the complexity of introducing semi-autonomous systems and the need to account for human factors and the potential misuse or overreliance on the technology.

Figure 5.7 A DMS ("driver surveillance" might not go down well with the marketing department!).

It is too soon to know what unintended consequences or misuse patterns might arise from the introduction of DMS. Drivers may develop a false sense of security or become overly complacent, relying too heavily on the system to monitor their attentiveness, leading to potential safety issues. There are also concerns about the privacy implications of continuous driver monitoring, the potential for data misuse, and hacking vulnerabilities. Adding DMS to vehicles will inevitably increase its cost, which could be a barrier to widespread adoption, particularly in more price-sensitive segments of the market. Like any technology, DMS may not always function perfectly, leading to potential false alarms or missed detections, which could undermine its effectiveness and erode consumer trust. As these technologies continue to evolve, it will be critical to strike the right balance between safety, convenience, personal freedoms, and cost-effectiveness, while also anticipating and mitigating potential unintended consequences.

What if ADAS was designed more broadly to eliminate deliberate driving maneuvers and behaviors that caused accidents? Imagine if vehicles had a brake light on the front as well as on the rear so that oncoming traffic could take evasive action, if necessary. Or what if the vehicle's forward-facing camera sensed a red light and automatically slowed down to prevent going through a red traffic light? Or if the vehicle speed was electronically governed when driving in low-speed areas inhabited by pedestrians and cyclists? If speed governing is too draconian, then perhaps the accelerator pedal could introduce some haptic resistance. In addition, because drunk driving and distracted driving are major causes of accidents, what if the vehicle could be prevented from starting if the driver's blood alcohol concentration was over the legal limit or if cell phone usage was blocked when the vehicle is being driven? Applying the sophisticated sensor fusion and AI capabilities that have been developed for ADAS to these challenges could create workable solutions that minimize false positives or negatives, and might reduce fatalities on the road far more effectively and with a lower cost than today's ADAS. These more proactive and preventive measures have often been discussed but never implemented because they are seen as too draconian, raising concerns about personal freedom, privacy, and the potential for mistakes, which would need to be carefully addressed. However, they would potentially tackle some of the root causes of accidents more directly, without the need for extensive driver monitoring or last-second autonomous interventions that create their own issues.

5.6. AVs

A feature like AEB, where the vehicle drives autonomously for a few seconds prior to an impending collision, illustrates the benefit of autonomous operation in space- and time-constrained conditions. For all practical purposes, a new AEB-equipped vehicle is manually driven, but it will take over control if a crash is imminent and if the driver is not responding appropriately. In fact, limited "last seconds" autonomous operation in a human-driven vehicle, and the countermeasures mentioned above drastically reduces the main benefit touted by AV advocates, which is safety.

5.6.1. Safety

According to the WHO, nearly 1.2 million people die each year from car accidents around the world, and it is the leading cause of death for people aged 5–29 years [5.22]. It should be noted that more than half of these deaths are vulnerable road users (pedestrians, cyclists, motorcyclists) and that the economic damage from traffic accidents is estimated to be around 3% for many developed countries. In fact, the fraction of people killed "inside the vehicle" (in passenger cars, light trucks, commercial vehicles, and buses) has declined consistently from 80% in 1996 to 64% in 2022, while the proportion of people killed "outside the vehicle" (motorcyclists, pedestrians, cyclists) has increased from a low of 20% in 1996 to a high of 36% in 2022 [5.23]. If vehicles become heavier, with large batteries, this trend could even accelerate.

Although safety is often mentioned as the main reason for developing autonomous passenger vehicles, there does not appear to be any available data that unambiguously proves how AV safety compares with an equivalent human driver, driving a similar vehicle in the same road and weather conditions. If the comparison is with a robotaxi, then the equivalent human driver should be a taxi driver who is presumably not drunk or distracted. Such data, if they exist, are being kept secret by ride-hailing companies like Uber and Lyft.

Or consider autonomous heavy-duty (Class 8) trucks that are being developed to address labor shortages and to drive almost nonstop, promising same-day delivery for goods shipped over the road from more than 1000 miles away. According to the Federal Motor Carrier Safety Administration (a branch of USDOT), there were 0.17 fatal crashes involving large trucks and buses per 100 million VMT in 2021 [5.24]. In other words, autonomous truck developers might have to drive 600 million VMT without a fatal crash to claim they are just as safe. To put this in perspective, a fleet of 100 autonomous trucks, driving 100,000 miles a year, might require 60 years of operation! And autonomous trucks are only operating on the highway and in good weather conditions, to accelerate deployment, which means that the actual safety comparison is likely tougher to meet because highway trucks must operate in all weather conditions and on surface streets as well as highways. A report, published in 2016 by RAND, concluded that for fatalities and injuries, test-driving by itself cannot be a practical method to prove AV safety to regulators and the public [5.25]. AV developers are stressing the important role that simulation can play in virtually testing countless variations of challenging scenarios faced in the real world, but ultimately, the challenge occurs when an unforeseen situation arises that has not been modeled in simulation.

We also know that the introduction of AVs can lead to new types of safety problems for other road users, as happened in San Francisco in

October 2023 when a pedestrian was run over by a human-driven car and thrown under the front of a Cruise robotaxi driving in the neighboring lane, making it a rare or edge-case scenario that may not have even been simulated. Robotaxis will need designated areas for picking up and dropping off passengers, which could create new conflict points with pedestrians, cyclists, and other vehicles if they obstruct traffic flow, block sight lines, or behave in an unpredictable manner. In some cases, AVs may need to be remotely operated or handed over to human control for certain edge-case situations, and the transition between autonomous and human operation could create new safety risks if not executed properly. Shared rides may pose greater personal safety risks for riders than with conventional taxis since there is no driver present, although the interior camera may be a deterrent. Or an autonomous truck might breakdown on the highway and move to the shoulder but be unable to place warning beacons 100 feet behind and in front to warn other drivers and prevent rear-end collisions. To mitigate some of these potential safety risks, several measures may be necessary, including regulations governing AV operation and interactions with other road users, clear communications, like external displays or signals, that convey the AV's intent and actions to surrounding road users, and rider education on designated waiting areas, for example.

Finally, as ADAS technology is integrated into vehicles, the capability exists to make human-driven vehicles much safer by preventing a distracted driver from drifting out of lane or crashing into the rear of the vehicle in front, as with the AEB feature that may be mandated for all new vehicles. In other words, autonomy can be enabled just prior to an imminent crash or dangerous maneuver instead of being engaged most or all of the time as a convenience feature. Realistically, AV safety should be compared not only with a trained driver (if it is a commercial vehicle such as a robotaxi or goods delivery truck) but with a conventional, manually driven ADAS-equipped vehicle that can take over autonomous control in an emergency, rather than the opposite of relying on the human driver as the backup!

5.6.2.
Accessibility

Providing enhanced mobility for the disabled and the elderly is a benefit touted by the AV industry. Even purpose-built, ground-up or "clean sheet" robotaxis, such as those made by Zoox, that have flat floors and wide door openings [5.26] are still not wheelchair-accessible since there is no ramp or wheelchair tie downs to attach it securely to the floor, as shown in **Figure 5.8**. In other words, in the US they are not being designed to meet Americans with Disability Act (ADA) standards. To be fair, Cruise had developed a wheelchair-accessible Origin shuttle before GM canceled its robotaxi initiative and Waymo does offer a wheelchair accessible vehicle (WAV) service, but robotaxis clearly lack a human "taxi" driver to assist the passenger in physically entering and exiting the vehicle, meaning that the wheelchair user still needs a personal assistant to help them. Inside the vehicle, the touchscreen interface or voice commands might not be accessible to users who have visual or cognitive disabilities. Even if the vehicle is suitably enabled, the passenger pickup and drop-off areas may not have ramps to bridge the gap between the sidewalk and the AV for wheelchair users.

Figure 5.8 A minivan designed for wheelchair users, with an automated ramp and with tie-downs to secure the wheelchair inside the minivan; bottom: Waymo's purpose-built minivan, developed by Zeekr.

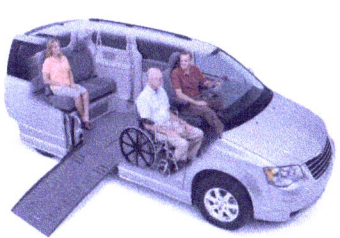
Image provided courtesy of BraunAbility®.

Image provided courtesy of BraunAbility®.

Karolis Kavolelis/Shutterstock.com.

Image provided courtesy of BraunAbility®.

Creating robotaxis with a "universal design" character that accommodates all users will add extra vehicle content. If, as an illustration, ADA-optimization added $3000 to the robotaxi vehicle cost if it were standard across the fleet and not a retrofit, then this would add one cent per mile to the robotaxi cost economics when spread over 300,000 miles of operation. The durability requirement for 300,000 miles seems a lot but the ADA-related components will not be used anywhere near as often as for a personally owned wheelchair accessible minivan retrofit. Since the average length of an Uber trip is around five miles then making each robotaxi meet WAV requirements might only add five cents, to the average ride, which is less than the "penalty" administered by, for example, NYC or California (as will be discussed in Chapter 7, Section 7.6). Robotaxi companies might have to decide whether to generate a "one size fits all" universal design for all users that may be more expensive, or to create a limited number of wheelchair-accessible AVs and manage the complexity of two types of vehicles in their fleet for different users.

5.6.3.
Congestion

For progressive cities that are promoting active modes of travel (e.g., walking, cycling) and aiming to increase public transport usage by adding more bicycle and bus lanes, robotaxis and autonomous goods delivery vehicles promise to add more pressure on the valuable curbside because they will need convenient pickup and drop-off points.

There is no consensus on the effect that AVs will have on congestion, public transport, and underserved populations, but what we do know is that if mobility becomes cheaper or easier, then it will induce demand for more mobility. This is clear to anyone who has seen a new road or lane created and then filled up almost immediately, and it is the principle behind trying to reverse it with congestion charging. The effects of TNCs on traffic have been studied and found to have increased congestion in those cities where they operate because they lower the cost of a taxi ride. Therefore, if robotaxis are significantly cheaper than owning a car or taking an Uber or Lyft, more people might use them instead of using alternatives (walking, cycling, public transport). This will lead to more vehicles on the road, potentially increasing congestion. Without passengers, "zombie" robotaxis might also cruise around searching for fares, adding further to traffic congestion.

On the other hand, AVs could communicate with each other and with traffic management systems to smooth traffic flow and optimize routes, reducing congestion and travel times. According to a 2019 University of Cambridge study, driverless cars could improve overall traffic flow by up to 35% if AVs coordinate with each other and platoon, where possible, to free up road space [5.27]. And with less need for parking, robotaxis could even help the city by freeing up valuable real estate. After all, parking for privately owned vehicles (parking lots and street parking) consumes up to 30% of the land area in the central business districts of major US cities and typically 10% of the total incorporated land for the whole metropolitan area.

On the "carrot" side, cities could create dedicated lanes, or use existing underutilized bicycle or bus lanes for robotaxis and/or automated shuttles to improve their efficiency and reduce their impact on general traffic. City planners might establish a "stick" such as a congestion fee or a limit on the amount of time a robotaxi can cruise around without a passenger to discourage unnecessary trips and incentivize efficient use, while plowing the fees back into improving public transport. Unlike robotaxis, autonomous trucks have the potential to reduce congestion because it is conceivable for them to shift daytime deliveries (that contribute to congestion, especially when delivery vans are double-parked) to overnight operation, when the road capacity is heavily underutilized. This time-shifting is clearly not practical for moving people in a robotaxi! A possible unintended consequence if there is an overall congestion charge (not only directed at AVs) is that it could actually stimulate more AV fleet operation, because the fee may be less than the gain in AV productivity caused by reduced congestion, allowing faster deliveries or more passenger trips. A future outcome of congestion charging could even be a shift in traffic from privately owned, manually driven vehicles to AV fleets, and it is conceivable that companies like Amazon and Alphabet might support cities and finance congestion charging initiatives because their urban AV fleets (autonomous goods delivery and robotaxis) stand to benefit.

5.6.4.
Energy Usage

Will AVs improve energy efficiency? That depends, to a large extent, on how much the compute requirements for autonomous operation can be reduced. The amount of power needed for high-performance computing systems in AVs is a moving target, but a reasonable estimate might be around 500 W of power to run the calculations on perception, object detection, path planning, and so on. Additional power is consumed by the autonomy sensors themselves and by the vehicle's cooling system for maintaining the compute platform's temperature, so that the total effect of autonomy might be to consume an average of 600 W of power. This means that an AV operating 10 hours a day might require up to 6 kWh to support the autonomy compute platform. During the same time, it might consume 50 kWh for driving 150 miles and another 10 kWh for accessories, such as air-conditioning. In other words, autonomy compute resources might consume 40 Wh/mile and could reduce energy efficiency by approximately 15%. On the positive side, AV operation can be designed to go no faster than the speed limit and can offer smoother, more controlled coasting and acceleration, and this could reduce energy usage by as much as 10% according to research performed in the US Department of Energy's NEXTCAR program (Next Generation Energy Technologies for Connected and Automated On-Road Vehicles) [5.28]. If it also leads to smoother traffic flow overall, then the fuel economy benefits also apply to multiple nearby vehicles. However, these benefits do not require AVs because, according to the results from the same program, outfitting vehicles with just ACC and coordinating them with V2X communications can provide 20% fuel economy benefits without incurring the significant energy consumption required for full autonomy. As with safety, it is instructive to compare the true efficiency benefits of AVs not with typical human-driven vehicles but, instead, with vehicles that are equipped with ADAS and V2X technology and can drive "manually" with cooperative ACC performance.

From an energy perspective, it is also important to account for the energy consumed by powerful computing resources in the cloud for training the algorithms in the first place and keeping them updated. A single two-megapixel camera (24 bits per pixel) operating at 30 fps generates 1440 MB of data every second, so even just two hours of operation per day can generate 1 TB of uncompressed data. Because robotaxis often contain more than ten cameras as well as multiple lidars and other sensors, it is easy for the uploaded data from a fleet of vehicles to become a staggering amount every day!

One Tesla cybertruck owner, who unwittingly agreed to share his vehicle's Full Self-Driving (FSD) data, has publicly stated that his vehicle has sent and received 940 GB of data in a month, compared to his wife's Tesla Model 3 that used only 2.5 GB of data during the same timeframe because she had disabled FSD-sharing [5.29]. Training a large language model (LLM) like ChatGPT-4 may consume more than 50 GWh of energy [5.30], and a reasonable assumption is that the energy consumed in training AVs may be roughly one-tenth of that for LLMs, and may never end as it will continue to be improved with usage and learning. Energy is also required to upload and transfer data across the cellular network to data centers, for storing the data and for downloading OTA updates, such as the latest version of FSD. Let us assume that 5 GWh of energy has already been used in training the software for the robotaxi developer and, because robotaxi commercialization has not happened yet, this one-time training energy cost will continue to grow. If this model is deployed at scale across a fleet of 1000 robotaxis for a large city, then the amortized energy consumption for software development for each robotaxi is 5 MWh. This seems a lot, but if the robotaxis operate for several years and each one drives around 300,000 miles, then the one-time training energy cost per robotaxi is effectively 17 Wh/mile in addition to the robotaxi's onboard 40 Wh/mile compute energy consumption. If a minivan consumes up to 400 Wh/mile for driving purposes then the on-board compute processing for autonomy may increase energy consumption by 10%. However, when the extra energy for training the autonomy algorithms in the cloud is also added then the overall impact could be that autonomy increases energy usage by more than 20% versus the "same" manually driven vehicle!

It is to be hoped that most of this training can be applied to new locations so that there is a scaling benefit but in addition to the "one-time" energy used in training the algorithms for autonomous driving, there is also ongoing energy consumed in processing, analyzing, and storing the data generated by the vehicles in the cloud. The resulting OTA updates also require preparation and distribution, and may also involve large software payloads, large-scale data centers, and cloud infrastructure, which can consume tens of megawatt hours of energy each day for computing, storage, and cooling.

Since gasoline or electricity production and battery material extraction energy requirements are included in well-to-wheels energy analysis of different propulsion systems, it becomes increasingly important to track true vehicle energy consumption for connected and increasingly AVs, which must include the energy used by cloud infrastructure, especially for power-hungry AI applications like autonomous driving.

A more glaring omission in the comparison between manual and autonomous driving energy consumption is that AVs are very likely to generate additional miles of travel, some of which may be zero occupancy. For instance, it could lead to more urban sprawl and longer, more energy-intensive commutes as people may choose to live in less expensive places and can use the additional time spent commuting in AVs more productively. On the positive side, this might allow people to afford housing while having a good-paying job. Or consider that a robotaxi may circle around a city center to pick up the next passenger, or an owner may summon their vehicle via "virtual valet" from a parking lot, or a highway autonomous truck may operate nearly around the clock, without a break, delivering goods. These scenarios reveal the types of additional induced energy that AVs will consume. The benefit associated with this unmanned or zero-occupancy autonomous operation is, effectively, the convenience of having goods delivered from further away on the same day or the convenience of not having to walk to one's parked vehicle.

In addition, there is the extra wear on the roads from the extra VMT, which will lead to more frequent repairs and environmental impact. And there is the extra consumption of packaged materials with their embedded energy and emissions, that were ordered online, stimulated by faster delivery. In short, the energy usage on and off the vehicle for enabling autonomous capability, and the induced demand for mobility caused by autonomous operation, is likely to dwarf any possible energy reduction caused by smoother vehicle operation and coordinated traffic (neither of which requires fully autonomous operation).

5.6.5.
Workforce and the Economy

One US study, published in 2017, has estimated that the economic gains from AVs could reach $936B per year because of savings from reduced accidents and increased productivity [5.31], while a more recent study indicated the potential for adding $214B to GDP in the US and creating 2.4 million new jobs [5.32]. If autonomy lowers the cost of shipping goods by truck, then the overall market for goods delivery could grow substantially and require more local delivery drivers, more than offsetting the job loss on the relatively controlled long-haul highway routes that will be driven autonomously. Moreover, there are likely to be new employment opportunities created in fleet management, cybersecurity and cloud services, EV charging infrastructure, and remote assistance providers.

However, several million driving professionals are at risk of displacement in the US alone, including truckers, taxi/ride-hail drivers, and transit workers [5.33]. A significant number of other workers will be disrupted in the auto ecosystem, including automotive repair, insurance, and refueling. This could be exacerbated by a simultaneous shift from personal vehicle ownership to mobility subscription models. Shared AVs can empower workforce participation for the disabled, elderly, and youth populations while AV operation can transform urban design and land use for the better, turning vacant parking lots into green spaces and housing for low-income residents. In short, the picture is mixed and there will be winners and losers.

The disruptive impacts of autonomy and intelligent mobility have the potential to catalyze sweeping economic shifts on par with past technological revolutions. However, this transition will likely occur more quickly than previous economic revolutions and with unprecedented scale and disruption that will affect multiple sectors and geographies. Proactive planning and investment will be critical to not just mitigate risks, but to ensure new opportunities benefit the environment and the economy in a way that is socially fair.

5.7.
Vehicle Affordability

Vehicles have become more sophisticated over time, for a variety of reasons. In addition to vehicle electrification, autonomy, and connectivity, new features include more safety-related content such as sophisticated airbags and ESC, as well as more comfortable interiors with higher-quality materials, heated seats, touchscreens, ambient lighting, and so on. Such improvements have made vehicles safer, cleaner, and more comfortable, and the cost has been offset by continuous improvements and cost reductions in other vehicle systems so that vehicle affordability has remained relatively constant over time.

With the exception of some luxury and sports cars, the US automotive business model has traditionally involved producing lots of vehicles and then incentivizing customers to buy them with rebates. Table 5.1 reveals that most automobiles have been priced and sold as a commodity, because the cost per pound is comparable with a hamburger! There are, of course, some exceptions for luxury and sports cars that can carry a brand premium. The US automotive business model has traditionally involved producing lots of vehicles and then incentivizing customers to buy them with rebates! **Table 5.1** also shows how a smartphone is roughly 100 times more expensive per unit weight than a car. This striking difference indicates that automobiles could become more expensive as it becomes more like a "smartphone on wheels." However, even if automotive electronics do operate in a tougher environment (shocks, ambient temperature, etc.) and must have redundant, fail-safe operation, unlike consumer electronics, the cost of an automobile will still have a large concentration of commodity materials (such as glass, rubber, steel) so it is never likely to be as "cost-dense" as a smartphone.

Table 5.1 Relative cost per weight for three common items.

	Price ($)	Weight (lb)	Price/weight ratio ($/kg)	Number of chips (semiconductors)
Automobile (Toyota RAV4)	35,000	3500	10	100
Hamburger	5	0.5	10	0
Smartphone (iPhone)	700	0.4	1750	20

During COVID-19, the auto industry faced a challenge in securing the semiconductors they needed for their vehicles to operate, partly because the semiconductor suppliers were prioritizing their smartphone customers. With less inventory due to chip shortages, the automakers were forced to focus on selling the more profitable vehicle models (typically high-trimmed full-size trucks, SUVs, and luxury cars), instead of affordable sedans. This new strategy increased their overall profitability and helped their stock performance. They also realized that Tesla was not affected as much because Tesla vehicles used fewer semiconductors, and relied more on software to extract vehicle performance from the underlying hardware. Even after the chip shortage eased, the automakers decided to maintain their new focus on offering fewer models but with higher prices, leading to average new car prices rising at a substantially higher rate than inflation in recent years.

The Internal Revenue Service (IRS) provides a useful surrogate for vehicle affordability because it reimburses businesses for mileage expenses associated with both owning and operating a vehicle, including financing, depreciation, insurance, maintenance, and fuel. The IRS updates this figure annually and it has risen from around $0.27/mile in 1990 to $0.70/mile in 2025. This mileage rate is compared with median household income and inflation and normalized to 1990 levels, as shown in **Figure 5.9**. The median income is probably a more suitable metric for affordability than average income, which can be skewed higher by very high wage earners.

This recent hike in new vehicle prices might be a long-term feature because as vehicles are transitioning toward having electric propulsion, offering higher levels of autonomy and requiring a clean-sheet electrical and software architecture, the cost to develop them has significantly increased both for the hardware (e.g., new batteries, motors, sensors, compute platforms, and electrical architectures) and for the software (in terms of growth and complexity). Import tariffs will also raise vehicle prices near-term although the long-term effect is not clear. The likely result is that new vehicle prices may continue to outpace inflation as well as median household income growth. This runaway price escalation represents a major threat to personal vehicle ownership and access to mobility.

Figure 5.9 Vehicle affordability in the US, normalized against inflation and median household income (1990–2025) [5.34, 5.35].

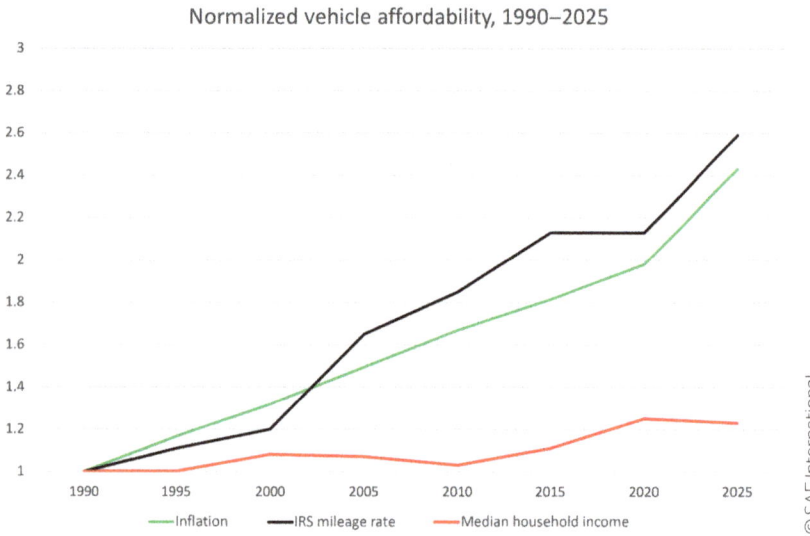

According to the Bureau of Labor Statistics and Kelley Blue Book, the average cost of vehicle repairs has also increased by 40% since 2012, a faster rate than inflation, and many more vehicles are now being totaled because repair is not economical. Auto insurance rates have grown 50% since 2021, also because of climate change, weather events, and a shortage of mechanics. Auto insurance now accounts for around 20% of the total cost of vehicle ownership, versus around 15% a few years ago [5.36].

One consequence of the worsening affordability of new vehicles is that people are keeping their cars longer before buying a new or used replacement. The average age of a car was just 5.6 years old in 1980 and increased to around 7.8 years in 1990. Now, as shown in **Figure 5.10**, the average age of US vehicles stands at 12.6 years, and it has been edging up each year by around a month [5.36]. S&P Global Mobility believes that around 110 million vehicles in the US, or 40% of the automobiles on the road, are at least this old [5.37].

Figure 5.10 Average age of vehicles in US in 2024 [5.37].

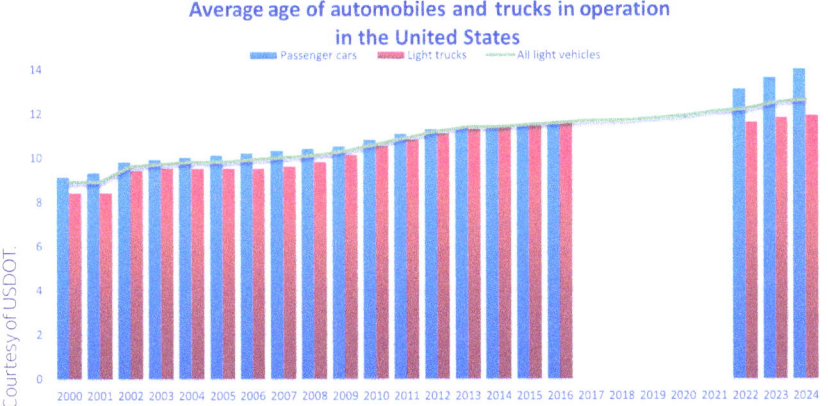

Because owning and driving the same car for 200,000 miles versus leasing a new car every three years can potentially save a consumer over $100,000, this trend may continue. As vehicles are being designed to last longer and have fewer moving parts that can fail, this may cause new vehicle sales to decline in the future and create challenges for automakers to find new revenue sources, such as software upgrades. Automakers have recently learnt how to become more profitable by following Tesla and the tech companies, and they are clearly beginning to focus more on software and looking to sell OTA updates, offering subscription services that unlock new features, and creating app stores for third-party developers. As existing automakers focus on wealthier customers, the rising cost of new vehicles creates an interesting opportunity for new competitors, perhaps from emerging economies.

References

5.1. Lienert, P., "When Do Electric Vehicles Become Cleaner than Gasoline Cars?", Reuters, July 7, 2021, https://www.reuters.com/business/autos-transportation/when-do-electric-vehicles-become-cleaner-than-gasoline-cars-2021-06-29/.

5.2. Sawal, R., "Red Seas and No Fish: Nickel Mining Takes Its Toll on Indonesia's Spice Islands," Mongabay, February 16, 2022, https://news.mongabay.com/2022/02/red-seas-and-no-fish-nickel-mining-takes-its-toll-on-indonesias-spice-islands/.

5.3. Bamana, G. et al., "Addressing the Social Life Cycle Inventory Analysis Data Gap: Insights from a Case Study of Cobalt Mining in the Democratic Republic of the Congo," *One Earth* 4, no. 12: 1704-1714, doi:https://doi.org/10.1016/j.oneear.2021.11.007.

5.4. Latief, Y., "Three Quarters of Global Battery Supply Chain at Risk of Labour Law Violations Says Infyos," Smart Energy International, September 18, 2024, https://www.smart-energy.com/industry-sectors/storage/three-quarters-of-global-battery-supply-chain-at-risk-of-labour-law-violations-says-infyos/.

5.5. Amnesty International Website, "Recharge for Rights," October 15, 2024, https://www.amnesty.org/en/latest/news/2024/10/human-rights-ranking-electric-vehicle-industry/.

5.6. Frauenhofer Research Institution for Battery Cell Production FFB, "Study on the Battery Supply Chain Shows China's Global Dominance - And Options for Europe," Press Release, February 18, 2025, https://www.ffb.fraunhofer.de/en/press/news/Chinas_Dominanz_in_der_Batterielieferkette.html.

5.7. IEA, "Recycling of Critical Materials," IEA Report, November 2024, https://www.iea.org/reports/recycling-of-critical-minerals.

5.8. "Glencore Plans Europe's Biggest Electric Car Battery Recycling Plant," Financial Times, May 8, 2023, https://www.ft.com/content/ab593cac-5f7e-4b70-904d-94cc0251aaeb.

5.9. Caltrider, J., Rykov, M., and MacDonald, Z., "After Researching Cars and Privacy, Here's What Keeps Us up at Night," Mozilla Foundation, September 6, 2023, https://foundation.mozilla.org/en/privacynotincluded/articles/after-researching-cars-and-privacy-heres-what-keeps-us-up-at-night/.

5.10. George, P., "The Ultimate 'Connected Car' Nightmare Is Playing Out in China," InsideEVs, September 2, 2024, https://insideevs.com/news/732178/car-software-fail-wm-motor/.

5.11. National Highway Traffic Safety Administration, "Critical Reasons for Crashes Investigated in the National Motor Vehicle Crash Causation Survey," February 2015, https://crashstats.nhtsa.dot.gov/Api/Public/ViewPublication/812115.

5.12. Schaff, S., "The War on Error? Rethinking What Matters in Automated Vehicle Safety," FrontierGroup, August 3, 2018, https://frontiergroup.org/articles/war-error-rethinking-what-matters-automated-vehicle-safety/.

5.13. Dong, Y. et al., "Evaluation of Crash Contributing Factors," *Journal of Transportation Technologies* 15, no. 1 (2025): 155-178.

5.14. AAA, "Cost of Advanced Driver Assistance Systems (ADAS) Repairs," December 2023, https://newsroom.aaa.com/wp-content/uploads/2023/11/Report_Cost-of-ADAS-Repairs-FINAL-23.pdf.

5.15. Cicchino, J., "Effects of Forward Collision Warning and Automatic Emergency Braking on Rear-End Crashes Involving Pickup Trucks," IIHS, https://www.iihs.org/topics/bibliography/ref/2265.

5.16. Sternlund, S., "The Safety Potential of Lane Departure Warning Systems—A Descriptive Real-World Study of Fatal Lane Departure Passenger car Crashes in Sweden," *Traffic Injury Prevention* 18, no. Supplement 1 (2017): S18-S23.

5.17. NHTSA, "Early Estimate of Motor Vehicle Traffic Fatalities for the First Quarter of 2024," June 2024, https://crashstats.nhtsa.dot.gov/Api/Public/ViewPublication/813598.

5.18. Gross, A., "AAA Finds Better Behavior behind the Wheel, But There's Room for Improvement," AAA, October 2021, https://newsroom.aaa.com/2021/10/aaa-finds-better-behavior-behind-the-wheel-but-theres-room-for-improvement/.

5.19. Hope, G., "Study Finds Drivers Distracted by Automated Technology," IOT World Today, September 18, 2024, https://www.iotworldtoday.com/transportation-logistics/study-finds-drivers-distracted-by-automated-technology#close-modal.

5.20. Large, D.R., Harvey, C., and Burnett, G., "How Will Drivers and Passengers Interact in Future Automated Vehicles?," Human Factors Research Group, University of Nottingham, June 2024, https://www.racfoundation.org/wp-content/uploads/How-will-drivers-and-passengers-interact-in-future-automated-vehicles-Large-et-al-July-2024.pdf.

5.21. Barry, K., "Clearing the Confusion about Advanced Car-Safety Feature Names," Consumer Reports, Updated January 25, 2024, https://www.consumerreports.org/cars/car-safety/clearing-confusion-about-advanced-car-safety-feature-names-a1035752654/.

5.22. World Health Organization, "Road Traffic Injuries," December 13, 2023, https://www.who.int/news-room/fact-sheets/detail/road-traffic-injuries#:~:text=Approximately%201.19%20million%20people%20die,adults%20aged%205%E2%80%9329%20years.

5.23. U.S. Department of Transportation, "Overview of Motor Vehicle Crashes in 2022," DOT HS 813 560, June 2024 (revised), https://crashstats.nhtsa.dot.gov/Api/Public/ViewPublication/813560.

5.24. Federal Motor Carrier Safety Administration, "Large Truck and Bus Facts 2021," https://www.fmcsa.dot.gov/safety/data-and-statistics/large-truck-and-bus-crash-facts-2021.

5.25. Kaira, N. and Paddock, S.M., "Driving to Safety: How Many Miles of Driving Would It Take to Demonstrate Autonomous Vehicle Reliability," RAND Corporation, April 12, 2016, https://www.rand.org/pubs/research_reports/RR1478.html.

5.26. Al-Heeti, A., "No Steering Wheel, Pedals or Driver's Seat: Is Zoox the Future of Robotaxis?," CNET, November 15, 2024, https://www.cnet.com/roadshow/news/no-steering-wheel-pedals-or-drivers-seat-is-zoox-the-future-of-robotaxis/.

5.27. "Driverless Cars Working Together Can Speed up Traffic by 35 Percent," University of Cambridge, May 19, 2019, https://www.sciencedaily.com/releases/2019/05/190519191641.htm.

5.28. Jones, W.D., "Autonomous Vehicles Can Make All Cars More Efficient", IEEE Spectrum, July 3, 2020, https://spectrum.ieee.org/autonomous-vehicles-fuel-efficiency?utm_source=thefuturelane&utm_medium=email&utm_campaign=thefuturelane-07-10-24&mkt_tok=NzU2LUdQSC04OTkAAAGUO_P8ubZk9iiPApV9HIqqIrUoqXgnW_1j4N9BN-ozu2FEcvzfRJI_LaSFgcTn4AVsE2PPT0z4q4eSzyu_fiJg_ab7K2NIwQ04B_cYPsRPbmM9NJ4.

5.29. Aregay, T., "A Tesla Cybertruck Owner Says He's Blowing Past His 1.28 Terabyte Data Cap after Buying the Truck – Adds, The Internet Spikes Coincide with Cybertruck Updates," Torque News, March 19, 2025, https://www.torquenews.com/11826/tesla-cybertruck-owner-says-hes-blowing-past-his-128-terabyte-data-cap-after-buying-truck.

5.30. Ludvigsen, K.G.A., "The Carbon footprint of GPT-4," Towards Data Science Website, July 18, 2023, https://towardsdatascience.com/the-carbon-footprint-of-gpt-4-d6c676eb21ae/.

5.31. Kockelman, K. et al., "Economic Effects of Autonomous Vehicles," *Transportation Research Record* 2606, no. 1 (2017), doi:https://journals.sagepub.com/doi/10.3141/2606-14.

5.32. Winston, C. and Karpilow, Q., *Autonomous Vehicles: The Road to Economic Growth?* (Brookings Institution, 2020), ISBN:9780815738572 (paperback book).

5.33. RethinkX, "How Many Jobs Will Be Created by the Disruption of Transportation? How Many Jobs Will Be Destroyed?," May 6, 2024, https://www.rethinkx.com/faq-and-mythbusting/how-many-jobs-will-be-created-by-the-disruption-of-transportation-how-many-jobs-will-be-destroyed.

5.34. "Median Household Income in the United States from 1990 to 2023 (in 2022 Dollars)," https://www.statista.com/statistics/200838/median-household-income-in-the-united-states/.

5.35. Wikipedia, "Business Mileage Reimbursement Rate," https://en.wikipedia.org/wiki/Business_mileage_reimbursement_rate.

5.36. Aeppel, T., "Soaring Insurance Costs Hit as US Buyers Get a Break on Car Prices," Reuters, April 11, 2024, https://www.reuters.com/markets/us/soaring-insurance-costs-hit-us-buyers-finally-get-break-car-prices-2024-04-11/.

5.37. Parekh, N. and Campau, T., "Average Age of New Vehicle Hits New Record in 2024," S&P Global Mobility, https://www.spglobal.com/mobility/en/research-analysis/average-age-vehicles-united-states-2024.html.

paradigm shift
changer
creativity
idea
knowledge
approach
shifting
vision
motivate
future
new paradigm
adapting
rethink
strategy
process
creative
assumptions
opportunity
disruptive
diverging
challenge
game
revolution
adaptation
evolution
fundamental
disruption
solution
different
improvement
innovate
shake
basic
paradigm
volutionary
radical
plan
drastic
proposal
conce
disrupt
ental change
rese
shift
leap
basic concept
innovative
status
change
scientif
evolve
inspiration
evolving
innovation
practice
upheaval
dissimilar
business
science
adapt
motivation
thinking
unique
disrupting
response
new
progress

Chapter | **06**

A New Paradigm for Auto-Mobility

Putting the heart back into technology clearly does not make literal sense because technology, as many people know, has always been used for both good and bad purposes. The difference is in the mindset of technologists, businesses, and governments. The previous chapter outlined some of the negative environmental, societal, and personal consequences that may occur with emerging automotive-driven technology applications of electrification, connectivity, and autonomy. However, the very same technologies and solutions can be applied in a much more beneficial way to support auto-mobility integration and to provide affordable and sustainable mobility for all. This automotive technology disruption is occurring at precisely the same time as the global automotive business landscape is being transformed and with existential challenges facing traditional automakers. Before proposing a new and comprehensive paradigm for auto-mobility, it is instructive to first understand how the global automotive business is likely to change in the coming years as this can create the opportunity and need for the new paradigm.

6.1.
The Automotive Technology Race

In the early twentieth century, the US was a major force in automotive innovation. Ford introduced mass production into the assembly line in 1913 and controlled around 90% of the US car market by the mid-1920s. However, they limited color choice, famously, to black because black paint dried more quickly, increased productivity, and lowered cost. This failure to innovate created an opening for more differentiation, and GM became the first automaker to promote a brand portfolio ("a car for every purse and purpose" was the slogan) with Chevrolet catering to entry-level buyers, Cadillac for the wealthy, and intermediate brands for middle-class professionals. Shortly afterward, in 1927, GM created the first styling department, which would usher in the annual model change to encourage new vehicle purchases. However, for the remainder of the twentieth century, despite America's global economic dominance, major automotive engineering innovations tended to emerge either from Germany's premium automotive sector (Mercedes-Benz, BMW, Audi, Bosch, and others) in areas like chassis systems and safety, or later on from Japan (e.g., Toyota and Honda) as they developed lean manufacturing processes to produce higher-quality and more reliable vehicles, and also developed more efficient engines, culminating in hybrid electric propulsion solutions. American automotive innovation, meanwhile, tended to focus on new vehicle types (e.g., minivans and SUVs) and comfort-related solutions (e.g., cruise control and air conditioning). As the auto industry was a self-contained industry segment, progress tended to be evolutionary and orderly, with suppliers selling technology solutions to mass-market vehicle brands only after an initial exclusive period with the premium vehicle manufacturers that had supported their technology development.

Then, around the time of the Financial Crisis in 2008, three companies, all based in Silicon Valley, began to disrupt the auto industry from different angles in ways that would reinforce each other, and the traditional automakers have been wrestling with the consequences ever since. It is not a coincidence that Waymo, Tesla, and Uber all started their efforts to reinvent autonomy, electrification, and shared mobility around the same time because advancements in computing power and key enabling technologies (lidar sensors, batteries, communications networks), growing concerns with mobility (safety, environment, congestion), and visionary leadership (Larry Page, Elon Musk, and Travis Kalanick, respectively) all combined with ready access to capital in Silicon Valley. This then stimulated further competition from other startups, Big Tech companies, and the established automakers and auto suppliers.

Compounding this technical challenge has been the economic one of competing with the simultaneous rise of the Chinese auto industry, particularly in the realm of EVs. Chinese automakers, such as BYD, Geely, and Li Auto, have made substantial investments in EV development and production, and tech giants like Baidu, Alibaba, and Tencent are actively pursuing AVs in similar ways to Alphabet and Amazon in the US. Having by far the world's largest automotive market and having substantial, strategic government support has meant that China is well positioned to be a fast mover in developing and scaling new automotive solutions for domestic use that can then be exported globally.

In summary, China and the US are now driving the future of the automobile at a rapid pace compared with historical auto industry innovation. Moreover, the enabling technologies are expected to be applied not only to personal vehicles but also to goods transportation, micromobility, and public transit systems. Examples include electric delivery vans, autonomous commercial highway trucks, electric scooters, roboshuttles, and eVTOLs.

6.2.
The "Inevitable" EV Transition

It is very likely that future vehicles, not just automobiles, will increasingly be EVs. This is due to a combination of factors such as cost reduction, performance improvement, vehicle choice, competition from China, energy security, and environmental concerns. Peak (ICEV) auto sales occurred in 2017, and as EVs grab a higher share of new vehicle sales, it is inevitable that the ICEV market shrinks rapidly, as is already being seen in China, and profits from ICEVs are likely to fall at an alarming rate for traditional automakers that are slow to transition to EVs [6.1].

The cost of lithium-ion batteries has been steadily declining because of performance improvements that reduce the amount of materials needed to store each unit of energy, as well as increased production that drives manufacturing efficiency and economies of scale. Moreover, new battery chemistries, such as sodium-ion (Na-ion), promise to offer even more choice and competition to help keep prices down. This is in stark contrast to the ICE, which has less opportunity for a technical breakthrough, is seeing less investment as automakers shift development funds to EVs, and consumes gasoline or diesel, for which there is no obvious alternative from an energy density or refueling infrastructure perspective. As battery costs continue to decrease, the upfront price gap between EVs and ICEVs will narrow and even disappear, making EVs more affordable than ICEVs and accessible to an even broader consumer base (**Figure 6.1**).

Figure 6.1 Annual global automotive sales and the shift toward EVs.

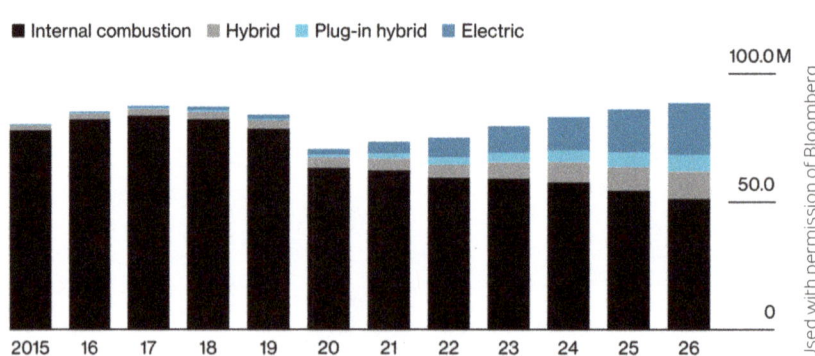

EVs are expected to become more affordable in terms of the upfront purchase price and already have lower operating costs, and their performance is expected to continually improve in terms of range, recharging time, and durability. Automakers will increasingly offer a wider choice of EVs, and they already tend to put their more advanced ADAS and infotainment technologies into these vehicles as current, relatively affluent EV buyers have a greater tendency to buy other technology features as well. As the EV charging infrastructure builds out and demonstrates proven availability, cost, and speed, the remaining barriers to consumer demand and acceptance should decline. There may be a small percentage of people who like the (genuine) sound of an engine, but this demographic is gradually shrinking because younger people in the developed world are more focused on access to mobility, and most people in the developing world who have never had a car do not share the same fondness or nostalgia for engine noise! It is expected that by 2035, the better performance, cost, and choice of EVs will attract most new car buyers, particularly in urban areas and in regions with incentives for EV adoption.

China's emergence as a global leader in EVs can be traced back to around 2007 when Wan Gang, a former Audi engineer, was appointed Minister of Science and Technology. Shortly afterward, in 2009, the Chinese Government began providing subsidies and tax breaks to EVs. China and its domestic automakers saw an opportunity to free themselves from relying on joint ventures (JVs) with foreign automakers that had superior ICE technology, while EVs also reduce dependence on Middle East oil and support climate change and local air quality initiatives.

In addition to helping secure access to battery raw materials from abroad, China's national government has taken a multipronged approach to supporting the development and commercialization of EVs and this includes coordinated support at the national, regional, and municipal levels. Between 2009 and 2023, the Center for Strategic and International Studies estimates that China invested $230.9B in building its EV industry [6.2]. This comprehensive support spanning consumer incentives, production mandates, infrastructure investments, and industrial development has been crucial in driving EV adoption and establishing China as a global leader in electric mobility and in becoming the world's largest EV market (**Table 6.1**).

Table 6.1 Chinese government policies supporting EV adoption.

National	Regional	Municipal
Mandated EV targets to drive EV production	Created EV pilot zones to test and showcase new policies and technologies	Restricted the registration of ICEVs to promote EV adoption
Provided R&D funding for domestic EV and battery technologies	Offered local subsidies and tax incentives	Provided free parking, access to bus lanes, and other benefits for EV owners
Supported the establishment of domestic EV and battery manufacturers	Designated specific industrial parks and development zones for EV and battery manufacturing	Partnered with local automakers to deploy EV fleets for public transportation and logistics
Invested in building a nationwide charging infrastructure network	Invested in building regional charging networks and battery swapping stations	Built extensive urban charging networks, including public charging stations and home/ workplace chargers

© SAE International.

To many people in the West, China's support of EV development may seem unfair, but even "laissez-faire" economies, such as the US, often protect, support, and invest in critical industries such as aerospace, pharmaceuticals, and semiconductors.

Chinese EV startups have the same kind of clean-sheet advantage that Tesla does when it comes to electric cars and software. Intense automotive competition in the US in the early twentieth century led to the rise of the Big Three (GM, Ford, and Chrysler) in the 1920s, and a similar dynamic in China today will lead to a few powerful Chinese automakers emerging and prospering domestically and abroad. Potential winners could include BYD (a battery company-turned automaker), Geely (who owns Volvo, London Taxi, and several other non-Chinese brands, as well as roughly 10% of Mercedes-Benz), and Xiaomi (better known for its smartphones). Even Chinese state-owned SAIC is attempting to transform itself by partnering with Huawei to jointly develop AI-enhanced EVs and cooperating in manufacturing, supply chain management, sales services, and in forming a joint brand, Shangjie [6.3]. China has the perfect environment to develop sophisticated automotive technology because it has ambitious and hard-working engineers and tech-savvy consumers. Automakers embrace creative solutions with less product liability, receive extensive government support, have an established and secure supply chain, and massive amounts of vehicle driving data for continuous learning and development.

EVs are inherently easier to develop and build because they do not require emissions calibration and have far fewer parts (perhaps 200 electric propulsion system parts, instead of well over 1000 for an ICE powertrain). As discussed in Chapter 5, Section 5.3.2, Chinese control of the battery materials supply chain is a concern shared by nearly all policymakers in the developed world because the auto industry is a major source of well-paid manufacturing jobs and a driver of economic growth. Several Chinese automakers, including BYD, Geely (and its premium Zeekr brand), and XPeng, are beginning to mass-produce affordable and attractive EVs for both domestic and international markets, and the intense competition from China is putting pressure on automakers worldwide to accelerate their EV and AI efforts in order to stay competitive or even survive! For example, GM's market share in China has dropped from around 14.9% in 2015 to 8.6% in 2023, with profits in China dropping by more than 75% [6.4]. Volkswagen has also been losing market share in China (19.3% in 2020 versus 14.5% in 2023) [6.5], and bought nearly 5% of Xpeng for $700M to access their software expertise. Toyota announced an R&D JV with BYD back in 2019 and a partnership with technology giant Tencent Holdings Ltd. in 2024 at the same time that Nissan signed a deal with another China tech giant, Baidu, on automotive AI.

It is likely that Chinese automakers will seek to form JVs in Europe and the US in order to gain market access and to establish a foothold in Western automotive markets. This is already happening as Chery is partnering with Ebro-EV (a Barcelona startup), and Stellantis has formed a JV with Leapmotor, a Chinese EV startup that is highly vertically integrated and develops its own battery packs, electric motors, electronic controllers, and even ADAS software. (In a nod to the smartphone, Leapmotor cars can even be started with a password!) The aim of the venture is for Leapmotor to export and sell its vehicles in Stellantis' European dealerships, with the JV splitting the profits, but the long-term vision is to make its vehicles in Stellantis' factories to avoid protectionist barriers.

However, the requirement to transfer EV technology as a condition for market access may face significant challenges and resistance, particularly in the US and certain European countries, because of intellectual property and national security concerns (even more so because EVs are also connected vehicles that can gather personal data and share this information with the Chinese government). A potential compromise might be for Chinese automakers to license technology or to establish R&D initiatives in Europe or the US that foster knowledge sharing while addressing intellectual property concerns. Such a compromise will need to balance market access, technology sharing, and strategic interests.

In addition to striving to remain competitive at home and abroad, particularly in the Chinese market, and with Chinese automakers globally, the traditional automakers are also having to respond to environmental regulatory pressures. Many countries and regions have established ambitious targets for reducing greenhouse gas emissions and promoting sustainable transportation solutions. In addition to climate change concerns, many European cities, in particular, have also created low-emission zones to reduce local air pollution. Given this competitive and regulatory background, it is not surprising that automakers continue to make pledges and commitments to phase out ICEVs and transition to full EV lineups over the coming decades. As emissions regulations tend to move in only one direction, automakers see the shift toward ZEVs as one way to simplify product development because the same EV can be sold everywhere, at least from an emissions perspective, because there are still some regional differences in crash safety regulations.

6.3.
How Have Countries with Major Automakers Responded to the Rise of EVs?

Between the time that Tesla "woke" the Chinese up to the possibilities of EVs and shook off dependence on ICE technology from foreign automakers (around 2010), and the time that the US, Europe, and Japan "woke" up to the threat of EVs from China (around 2020), several factors contributed to the developed countries falling behind China in EV development and adoption. These included a failure to invest adequately in long-term EV ecosystem solutions and short-sighted government support for their domestic automakers.

Compared to China's government, the US government did not invest as heavily in battery R&D, nor did it focus as much on securing supply chains for the critical raw materials that are needed for EV batteries. The US offered fewer incentives to EV and battery manufacturers compared to China and did not invest as heavily in charging infrastructure as China did. One can also argue that automakers fought for fuel economy standards that supported the continued development and production of ICEVs, potentially slowing down their EV development efforts. Moreover, the 2008–2009 auto industry bailouts of GM and Chrysler came without a strong EV mandate, and subsequent tax benefits for larger vehicles have indirectly supported ICEV production.

The Japanese government, meanwhile, has focused more on HEVs, because Toyota is a global leader and the dominant automaker in Japan with a 33% market share in 2023 (roughly as much as Honda and Nissan combined). Japan has also invested heavily in hydrogen fuel cell technology, supported by Toyota and Honda, which has probably diverted resources from EV development. Their support for small, efficient ICEVs (Kei cars) also potentially slowed EV market growth.

The EU, on the other hand, did implement increasingly stringent CO_2 emissions standards, indirectly pushing automakers toward EVs, and many European countries, such as Norway, offered significant tax breaks and subsidies for EV purchases. Many European cities have invested in electric buses and implemented low-emission zones and other policies favoring EVs. However, some European countries promoted diesel engines as a "cleaner" or more efficient alternative, delaying the shift to EVs, and often protected their domestic auto industries, which were heavily invested in ICE technology.

The common policies across these three regions (the US, Europe, and Japan) during the 2010–2020 decade included job protection in the traditional auto industry, which slowed the transition to EV production, and incentivizing consumer demand for larger, more powerful vehicles with ICE technology while government subsidies kept fossil fuel prices artificially low, making EVs less economically attractive.

The decade 2010–2020 was a "lost" decade for the traditional automaking nations, but since 2020, Europe, the US, and the rest of Asia (excluding China) have significantly increased or attracted investment in EV and battery manufacturing as the importance of China's leadership has sunk in. For example, because the Inflation Reduction Act was passed in the US in 2022, over $50B has been announced for battery manufacturing and more than $170B for EV manufacturing in the US in public and private investment [6.6]. However, this investment is just one piece of the puzzle for catching up because there still needs to be good access to the critical raw materials upstream for making batteries to use the invested manufacturing capacity, and there also needs to be adequate charging infrastructure investments downstream to increase EV adoption and help make EV manufacturing profitable (**Figure 6.2**).

Figure 6.2 Cumulative investment commitments by EV and EV battery manufacturers, by region.

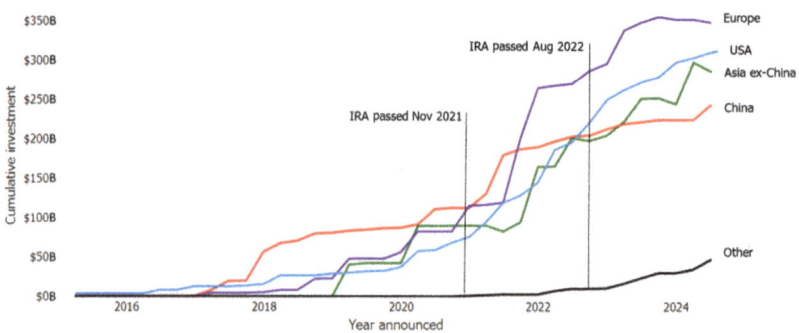

Atlas Public Policy, U.S. Investments in Electric Vehicle Manufacturing (2024), 2024 [Online]. Available: https://atlaspolicy.com/u-s-investments-in-electric-vehicle-manufacturing-2024.

6.4.
Emerging Countries in the EV Value Chain

As if the competition from China was not enough, several developing nations are sensing an opportunity to capture value, increase energy security, improve the environment, and develop and manufacture EVs as the auto industry shifts from a reliance on petroleum toward battery minerals, and from complex ICE technology to simpler electric drive. The two role models, particularly for mineral-rich nations, are China and OPEC.

Australia produces over half of the world's lithium, which is the key ingredient in lithium-ion batteries, and it is also a major supplier of other critical minerals like cobalt and nickel. The Australian government has identified the mining and processing of critical minerals as a strategic national priority to support the global energy transition and is investing heavily in building out its critical minerals supply chain, including exploration, mining, processing, and refining capabilities. The government has introduced incentives and support programs to attract investment in critical minerals projects and processing facilities inside Australia, and it is exploring opportunities to partner with other countries to develop integrated critical minerals supply chains.

Chile has vast copper reserves and low-cost mining operations, which make it the world's largest producer of copper, which is essential for the electrical wiring and motors in all types of vehicles, particularly EVs. The Chilean government is also nationalizing and exerting greater control over its vast lithium reserves, which account for more than 50% of the world's known lithium resources. It is investing in domestic lithium processing capabilities to move up the value chain, rather than just exporting raw lithium ore, and is partnering with foreign companies to develop lithium projects and lithium-ion battery manufacturing facilities inside Chile to increase its revenue from lithium exports (**Figure 6.3**). It has a good basis because the Chilean company SQM (Sociedad Quimica y Minera de Chile) is one of the largest lithium producers in the world.

Figure 6.3 Major sources for lithium-ion battery EV materials.

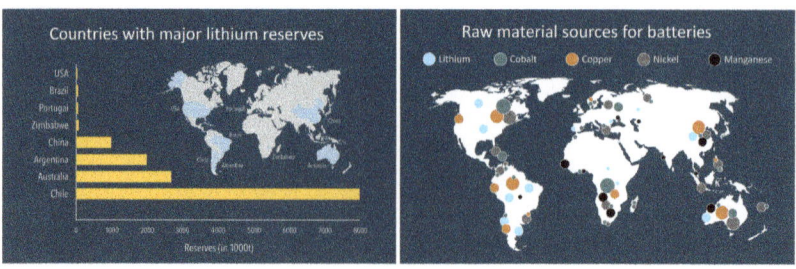

Dimitrios Karamitros/Shutterstock.com.   Dimitrios Karamitros/Shutterstock.com.

Chile and Australia have some of the most proactive and coordinated approaches to developing, controlling, and processing critical raw materials for the EV industry. Their focus on building domestic processing capabilities and securing strategic control over their mineral resources have positioned them to potentially capture a larger share of the value from the global EV supply chain. However, other countries are also looking to go up the EV materials value chain. Argentina and Bolivia, part of the Lithium Triangle with Chile, are also aiming to capture more value from lithium extraction and processing, although it will not be easy because of environmental and technical challenges. Both countries are trying to develop their lithium industry and attract investment in battery production to become a key player in the global lithium supply chain. Elsewhere in South America, Brazil is hoping to exploit its niobium reserves for powerful permanent magnets to be used in electric motors.

Indonesia has significant reserves of nickel, another key component in lithium-ion batteries (both NMC and NCA types), see **Figure 6.4**. It is implementing policies to encourage nickel processing and battery manufacturing within its borders. For example, it bans the export of raw nickel ore, has formed the Indonesia Battery Corporation with a 140 GWh battery cell production target by 2030, and has signed agreements with Chinese and South Korean battery and automobile companies for battery production. Indonesia also wants to capture more value by producing battery precursor materials (such as cathodes and anodes). Becoming an EV manufacturing base, however, may be challenging because the EVs that are more affordable to its domestic population use

LFP batteries that contain no nickel, meaning that EVs produced may need to be exported unless the Indonesian Government subsidizes them. Moreover, much of the EV-related foreign investment that has come into Indonesia has come from Chinese firms who assemble EVs in Indonesia from imported CKD kits, rather than setting up manufacturing plants that would help train Indonesia's workforce [6.7].

Figure 6.4 Nickel mining in Indonesia.

In Africa, the Democratic Republic of the Congo (DRC) supplies more than 70% of the world's cobalt, a critical component in lithium-ion NMC batteries used in EVs. The DRC is increasingly aware of its geopolitical importance and has been exploring ways to enhance its position in the global cobalt market, including through state-owned enterprises and JVs. However, the DRC poses significant supply chain risks because of political instability, poverty, and ethical concerns around mining practices. Zambia has significant supplies of copper and cobalt, making it another developing country that is positioned to capitalize on the EV supply chain growth. Neighboring Zimbabwe is also rich in lithium, nickel, and platinum group metals and is seeking to develop its mining sector and attract investment in battery material processing and refining. Like other countries, it is aiming to move beyond raw material exports and capture a larger share of the value chain.

A major producer of platinum group metals, South Africa is investing in battery technology and EV manufacturing and is also exploring opportunities to create a domestic battery value chain.

While none of these countries are forming a formal cartel like OPEC, their actions collectively aim to strengthen their negotiating positions and to secure a larger share of the profits generated by the burgeoning EV industry. Their policies generally involve nationalization or control over critical mineral resources, restricting exports of raw materials to ensure prices remain high, developing domestic industries for refining and processing the materials, and collaborating with foreign companies to accelerate technology transfer and build production capacity.

These countries and others possess abundant reserves of the critical minerals required for EV batteries and other clean energy technologies. By ramping up their mining and processing capabilities, they can generate substantial economic benefits in the form of export earnings, tax revenues, and job creation. However, it will be critical that the development of these mineral resources is done in a socially and environmentally responsible manner to ensure the local populations and ecosystems also benefit. These countries are critical to the EV battery supply chain, but each faces unique challenges in terms of scaling production and addressing environmental and socioeconomic, or fairness, concerns. Moreover, unlike oil and the existing gasoline distribution infrastructure, there are many alternative materials that can be used for future EV batteries. If, for example, cobalt or nickel prices or availability becomes an issue for automakers, then LFP batteries can often substitute for NMC types, and even if lithium supply is problematic, there may be Na-ion batteries that can be used in its place. This competition should limit the power that materials producers have over governments and automakers, and so the parallels with OPEC are not exact.

Going downstream in the EV value chain, countries such as Vietnam, India, and Saudi Arabia are trying to learn from China's example and positioning themselves to be engineering and manufacturing hubs for EVs that are relatively easy to design and build (compared with ICEVs). In the case of India, which already has a significant auto industry (JLR, for example, is owned by India's Tata Motors), the aim is to focus on

E2W/E3W and affordable EVs (passenger and commercial) that are well suited to local conditions with a view to exporting to nearby countries in the near-term and developed countries after that. Vietnam has long been a center for motorbikes, but with the emergence of simpler EVs, there may be an opportunity for one of its companies, VinFast, to become a significant manufacturer of affordable, high-quality EVs as well (**Figure 6.5**). The Vietnamese Government is supporting VinFast by, for example, subsidizing electricity at the VinFast charging stations, but the firm is currently struggling with quality issues, losing money, and staying afloat only with assistance from its billionaire owner [6.7].

Figure 6.5 VinFast car launch in Pasay, Philippines.

Saudi Arabia, meanwhile, is pursuing a broad strategy when it comes to financing and developing EVs, both within Saudi Arabia and abroad. In 2021, Saudi Arabia announced plans to invest $110B to develop a domestic EV manufacturing industry by 2030. This money would go toward EV manufacturing, battery manufacturing, and a nationwide charging infrastructure and would position Saudi Arabia as a Middle

East hub for EV production and sales. At the same time, Saudi Arabia's sovereign wealth fund is also actively investing in EVs abroad through investments in Lucid Motors, Nio, and Xpeng (**Figure 6.6**). This allows Saudi Arabia to gain exposure to cutting-edge EV technologies and manufacturing capabilities that are developed in other countries and to potentially export EVs produced by these companies to the Middle East.

Figure 6.6 Lucid's first factory, located in Casa Grande, Arizona. Their second factory is in Jeddah, Saudi Arabia, and is the country's first automotive manufacturing facility.

The United Arab Emirates (UAE) and Qatar are following a similar path, investing wealth generated from petroleum into renewable energy and electric mobility. For example, the UAE has partnered with a Chinese automotive company NWTN (pronounced "Newton") to establish an EV manufacturing facility in Abu Dhabi, and Qatar's Investment Authority has made significant investments in QuantumScape, a California-based company that is developing solid-state lithium-metal batteries for EVs and has also received investment from VW.

6.5.
How Did Traditional Automakers "Drop the Ball"?

Over the last 20 years or so, traditional automakers can claim to have made significant progress in advancing the automobile and their operations. For example, as discussed in Chapter 2, Section 2.1.2 similar-sized automobiles are more fuel-efficient and increasingly electric. Vehicles protect their occupants better and last longer. They have better infotainment and connectivity. The industry has embraced advanced robotics and lean manufacturing, globalization of their operations (particularly with investments in China), powertrain diversification, and flexible vehicle architectures. While they may not usually be at the cutting edge, traditional automakers have shown resilience and adaptability in many areas, and many are now positioning themselves for the future by investing heavily in electrification, autonomous driving, and new mobility services.

But along the way, many automakers have made notable mistakes. A prime example is the slow adoption of EV technology, initially allowing Tesla to gain a significant market share and then allowing China to dictate the future of the auto industry. The automakers underestimated the potential for growth of the EV market and did not invest sufficiently in battery technology or in creating compelling ground-up EVs, either in their home markets or in China. They relied on others to build out an EV charging infrastructure that was not being maintained in good condition and was plagued with issues for EV users, in stark contrast with Tesla's Supercharger network.

For example, at around the same time as Tesla was introducing its first new models, in 2008, Volkswagen was installing software as a "defeat device" to activate diesel engine pollution controls when it sensed an emissions test was taking place but to reduce those controls during normal driving, to protect engine components. VW's deception was not discovered until 2014 and as part of a settlement with the California Air Resources Board (CARB) and the Environmental Protection Agency (EPA), Volkswagen agreed to invest $2B in supporting EV charging infrastructure rollout in the US. The Detroit three have also made product decisions that were not aimed at reducing energy consumption or air pollution but actually moved in the opposite direction. By retreating from foreign markets where cars are more common, they ceased the development of most cars and increasingly relied on large SUVs and trucks in the US to generate profit and revenue. These larger vehicles are the least attractive to electrify because they need expensive and heavy battery packs to provide adequate range. If the price of energy (gasoline, electricity, and batteries) suddenly spikes in the future, they will be vulnerable to a consumer preference for more efficient vehicles, as happened in the 1970s during the oil crisis. Meanwhile, Toyota's brand image and leadership in HEVs, coupled with their sincere belief that this is a better environmental solution, has arguably held them back from investing more in EVs.

Once the threat from Tesla and China became clear, traditional automakers did decide to invest heavily in EVs. For example, Ford and GM have each committed around $60B to EV and battery manufacturing in the US while even non-US automakers like Toyota and VW have invested over $10B in the US [6.6]. However, unlike EV startups, these automakers are at a disadvantage because they are not pure EV players, which means they must balance a continued investment in ICE technology, hybrids, plug-in hybrids, and even hydrogen fuel cell vehicles. This juggling act makes a commitment to pure EV strategies far more difficult to maintain under ever-changing circumstances.

Along with this on-off commitment to EVs, traditional automakers may not have invested adequately in software development. This has meant that they are falling behind in developing advanced infotainment systems, producing vehicles that have many software issues and recalls, and with a limited capability to update OTA.

Despite their involvement in DARPA's Urban Challenge, and even sponsoring the winning teams, automakers were forced to watch as tech companies hired the experts and took the lead in developing fully autonomous or L4 vehicles. In contrast, L2 and L3 autonomy occur in less complex driving conditions and can be handled by automakers and their existing suppliers with more conventional software approaches. GM/Honda, Ford/VW, and Hyundai either acquired or invested in L4

AV robotaxi software startups, but the vast amounts of funding and time required to bring L4 autonomy to market has proven too much for the automakers to stay in the game, and so they have each unwound their L4 activities. In a broader sense, automakers are struggling to hire sufficient software engineers to cope with the increased demands of SDVs, and partnerships between automakers and suppliers, despite having many technical and commercial challenges, are slowly emerging out of necessity in order to compete with nimbler startups and tech firms that can attract software talent more easily. Organizationally, decision makers within the traditional automakers still tend to have a mechanical engineering or finance background, but from a technical perspective, this has to change, and quickly.

Beyond electrification and autonomy, automakers have generally failed to develop products that are truly tailored to emerging customer needs, and over the last 20 years there have been no truly innovative vehicle designs like the 1980s minivan, the 1990s SUV, and the 2000s crossover (car-based "SUV" with less off-road capability but better fuel economy). Vehicles are looking more homogeneous and indistinguishable from each other today as the hatchback or crossover vehicle becomes the dominant vehicle design.

In the mobility realm, automakers have made half-hearted attempts to explore mobility services and carsharing fleets and platforms, but they have shown limited patience for supporting the scaling costs. Examples of short-lived initiatives include GM's Maven program (2016–2020) and BMW's partnership with Mercedes-Benz, which merged their respective DriveNow and Car2Go programs to form Share Now (2019–2022). In general, automakers have tended to treat these fleet sales as "inferior" to consumer sales because fleet vehicles often have lower profit margins and may be sold as taxis or rental cars, which can hurt the brand's image. Moreover, when fleet vehicles are sold *en masse* it can lower the resale value and affect the brand. However, Ford has recently begun to realize that its fleet division, Ford Pro, is highly profitable and that electrification, connectivity, and autonomy trends can be monetized more easily with fleet sales than with consumer sales. Fleet vehicle sales are also likely to grow at a faster pace in the future than personal vehicle sales. Increased goods delivery and robotaxi deployments are two reasons, and increased new vehicle prices may be another.

In terms of operational efficiency, the automakers' vulnerability to critical semiconductors was exposed during COVID-19 when they were forced to park lots of assembled vehicles, and when billions of dollars in inventory waited for months until the necessary semiconductors arrived and could be "plugged in," powering the vehicles to life and making them ready for sale. EV startups, working with clean-sheet electrical architectures, using fewer chips and relying more on software, were subsequently less affected by the chip shortage. These mistakes and others have left many traditional automakers playing catchup in key areas like electrification, autonomy, and infotainment with Tesla and many Chinese automakers.

Making matters even worse for employment in the traditional auto manufacturing countries is the fact that automakers have been using China as a base to produce and export vehicles to the rest of the world. For example, Tesla shipped 344,000 China-built cars to Canada, Australia, Europe, and other markets in 2023 [6.8], while GM sent tens of thousands of Chevys to Mexico and other markets, and Ford has exported more than 100,000 trucks and SUVs from China to Southeast Asia, Africa, and the Middle East. European automakers, such as Mercedes-Benz, Volkswagen, BMW, and Renault, and Asia's Hyundai, Nissan, and Honda are also exporting cars made in China to their home markets.

6.6.
Traditional Automakers Need a New, "Sustainable" Strategy

The early automobiles cost a few thousand dollars and could only be afforded by wealthy people. These vehicles were hand-crafted, were made in low volume, and used expensive materials. The mass production or assembly line approach, pioneered by Henry Ford, made cars less expensive to make and prices fell below $1000 with the Model T's introduction in 1908. Over time, the price would fall further in real terms as volumes increased and with limited factory options. Although aftermarket changes were commonplace, the assembly line was designed to make the work more repetitive and standardized, but this monotony also led to high employee turnover and was one of the reasons why Henry Ford introduced a $5/day minimum wage in 1914, more than doubling the established daily labor rate. By 1925, a basic Ford Model T cost $260 or two months of income for a Ford worker, quite affordable by modern standards and a key factor in the Model T's success. As other automakers in the US and abroad copied Ford, an automobile purchase became realistic even for working-class people, and this ultimately led to extensive motorization and suburbanization over time.

Ford's reluctance to produce a wide variety of vehicles led Alfred Sloan, GM's President at the time, to create a range of different automobiles that could be marketed to different types of consumers or workers. The Chevrolet brand pitched its products to entry-level or lower-income workers while Cadillacs were sold as luxury vehicles to wealthy consumers, and several other brands fitted in between. The idea was to segment the market and sell more vehicles with a higher average profit than was possible with a single vehicle. This approach was soon adopted by other automakers around the world. As the profits were disproportionately higher with high-end vehicles, it effectively enabled cheaper cars to be sold with a "subsidy" from the more expensive automobiles made by the same manufacturer.

However, as discussed in Chapter 5, Section 5.7, a new automobile purchase is increasingly out of reach for the working class and even the middle class. For example, an annual income of $250,000 is needed today for someone to afford an average new car with two months of income. Even the cheapest new car for sale in the US in 2024 requires an annual income exceeding $100,000 to afford with two months of income [6.9].

In addition to reframing how automotive integrates with mobility (auto-mobility), I will argue that automakers now have an opportunity to reframe the vehicle itself and stay relevant, putting all of the "world on wheels" again but in a sustainable manner. Traditional OEMs are "under attack" from a variety of forces, but this new strategy could even put them on the offensive, opening up new markets for their core engineering competency, because they do not currently sell many vehicles to people who have little income or who live in densely populated cities anyway, and they do not make other types of road vehicles.

An increasing demand for goods and services (cargo vans and planes), a greater emphasis on public transport systems to reduce traffic congestion and pollution (buses and trains), and the growth of ride-hailing platforms (cars) all point to a future where personally owned automobiles may no longer be the dominant source of road transport energy consumption and greenhouse gas emissions.

Change is often most effective when a company's survival is at risk, and in this case, the change needed aligns with the broader needs of society. Instead of being driven by technology disruptions, mobility solutions should aim to pragmatically address fundamental human needs, overcoming limitations that the alternatives cannot adequately serve, and with a balanced transportation ecosystem providing choices for everyone. There is also a human need for people and communities to be empowered and self-sufficient, which means that a different approach may be needed, one that looks at mobility as part of a civic community system, steps back, and asks the question "What do we want mobility for, and what can it do for us if we re-think it?"

A new framework for sustainable and affordable auto-mobility for all has a hierarchy that should begin with the goal of reducing the need for mobility in the first place, by allowing people to access other people, places, and things without needing motorized transportation. This is clearly impossible to achieve, especially in rural areas, but for urban areas, where around 70% of the world's people are expected to live in 2050, access can be helped with sensible policies regarding urban design.

6.7.
Hierarchy of the New Framework

Urban planning is the foundation for sustainable mobility because it prioritizes land use and transportation systems that minimize reliance on personally owned automobiles, designing places for people instead of for cars as has often been the case. Effective urban planning promotes mixed-use developments that encourage walking, cycling, and the use of public transport, thereby reducing traffic congestion and emissions. This approach results in healthier lifestyles because of increased exercise and reduced air pollution, and it enables more affordable access to employment, healthcare, education, and recreation opportunities. By designing cities that are more accessible, planners can enhance the quality of life for residents while promoting sustainable transport options.

Active mobility, which includes walking and cycling, is essential for reducing carbon footprints and improving public health (**Figure 6.7**). Cities that prioritize an active mobility infrastructure, such as bike lanes, bike sharing, and pedestrian pathways, not only reduce reliance on fossil fuels but also foster a sense of community and well-being. This shift toward active modes of transport can be supported by urban designs that encourage shorter travel distances and safer routes.

Figure 6.7 Cities with a high quality of life tend to promote active mobility and have diverse mobility options.

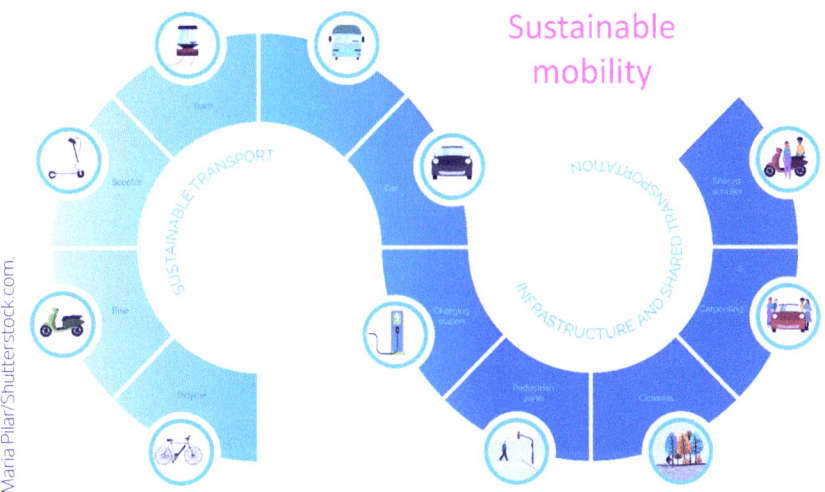

The next building block is to integrate various transportation services into a single accessible platform, allowing users to plan, book, and pay for multiple modes of transport seamlessly. If this MaaS approach is made seamless and convenient, it can encourage some drivers to leave their cars at home, thus reducing the number of vehicles on the road. By making public transport and shared mobility options more convenient, MaaS can help lower transportation costs for users and can be more convenient than driving a car because parking a car can add significant travel time and stress for drivers in crowded city centers. Shifting transport from personal vehicles to MaaS can also support local, regional, and national air quality initiatives (**Figure 6.8**).

Figure 6.8 A simplified chart showing the pros and cons of different forms of mobility.

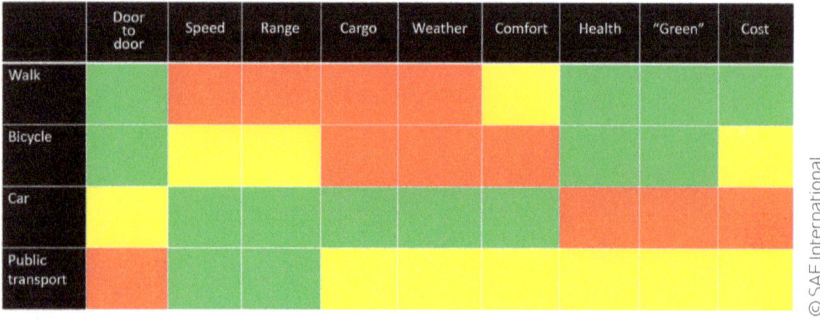

Before rushing to "ban" personal cars, we must acknowledge the value and flexibility they bring and the limitations that other modes of transport still have in addressing certain needs or situations. For example, walking or cycling is a healthy way to travel for short journeys, but it may not be practical for people with disabilities or who need to carry heavy, bulky items or who want to stay protected from the elements (wind, rain, snow, etc.), and it has obvious shortcomings for long trips. Public transport, where it exists, can address some of these limitations, and it can persuade many urban dwellers to forego owning a vehicle and the substantial, associated costs of purchasing, insuring, maintaining, and parking it. However, it is typically restricted to certain routes and schedules with frequent passenger stops, and, without MaaS, it does not go door-to-door, which means the person still needs to get to/from the public transport stop. There may also be safety and security concerns with being on a bus or train, or in waiting for one, particularly at night. Public transport operations can be disrupted or canceled by harsh weather conditions and, as with walking or cycling, there are limits to what can be carried or transported; for purchasing large items in the store and for hauling recreational equipment (bikes, skis, etc.), a vehicle with sufficient interior space and payload capacity can come in very handy. These limitations point to why using a vehicle may be needed but it does not mean it has to be owned.

With an advanced MaaS solution, it may be possible at some point to subscribe to a service for accessing cars, vans, and trucks on demand as easily as it is to access shared bicycles and scooters today. This creates a compelling argument for right-sizing vehicle design because many people currently purchase a vehicle that is grossly overengineered for, perhaps, 95% of trips. A right-sized vehicle for the vast majority of trips combined with convenient, seamless access to more capable vehicles "on demand" could make sense if the "friction" can be removed from the rental process, as will be discussed in Chapter 7, Section 7.4. Or someone may choose to not even own a vehicle but to use MaaS, including a subscription service to vehicles, all the time. The key to achieving this vision is to remove the pain points associated with vehicle rentals, including vehicle pickup and drop-off, insurance, and cleaning and maintenance. Integration of these "right-sized" vehicles with public transport can reduce traffic congestion and improve overall transportation efficiency, provide more sustainable and cost-effective mobility options for urban areas, and align with the trend of developing multimodal transportation systems.

The remaining elements of the new auto-mobility framework are focused on developing EVs to be more sustainable and affordable while accelerating a shift toward local production and community engagement. This begins with right-sizing to support batteries and vehicles that are appropriately sized for their typical, intended use. For instance, electric bikes and micro-EVs can serve urban environments more efficiently than traditional cars. This approach matches vehicle capacity to actual usage needs and reduces energy consumption, parking space requirements, and accident frequency and severity, while simultaneously making mobility more affordable by lowering operational costs.

In addition to right-sizing and light-weighting the vehicle to reduce energy demand, energy harvesting techniques may also be used to recover and harness energy from the vehicle and from the environment, and this will reduce energy needs even further. Examples of energy harvesting include regenerative braking (conversion of vehicle's mechanical energy, when slowing down, into electricity) and roof-mounted solar panels (conversion of sunlight into electricity). In both cases, the electricity generated can be fed back into the vehicle's battery to extend the range "for free." Of course, solar panels can also be integrated into the public transport infrastructure to power EVs and help cities become less dependent on external energy sources and to promote the use of renewable energy.

In addition to using less energy and renewable energy, EVs can be made more sustainable if they use materials that are either recycled or natural. The circular economy emphasizes the reuse and recycling of waste materials while using natural materials in the vehicle's construction can also enhance sustainability. In both cases, materials recovery and supply can be obtained locally and further support local economies so that this approach not only reduces the carbon footprint of mobility solutions but also fosters resilience in urban environments. As micro-EVs are easier to design and engineer than traditional automobiles, with much simpler requirements, they can use a wider variety of materials. Local micro-factories can be used to produce vehicle parts and EVs closer to where they will be used for moving people and goods. This reduces transportation emissions and supports local economies by creating even more jobs. By decentralizing production, communities can also tailor or customize mobility solutions to their specific needs, enhancing accessibility and affordability (**Figure 6.9**).

Government policies will play a pivotal role in establishing this new framework for sustainable and affordable auto-mobility. By providing incentives for EV production, investing in MaaS, and putting some constraints on private vehicle use, local or municipal governments can create an environment conducive to local manufacture and deployment of right-sized vehicles. Policies that promote the circular economy and local production can further enhance the sustainability of urban mobility systems.

Figure 6.9 Framework for sustainable and affordable mobility for all.

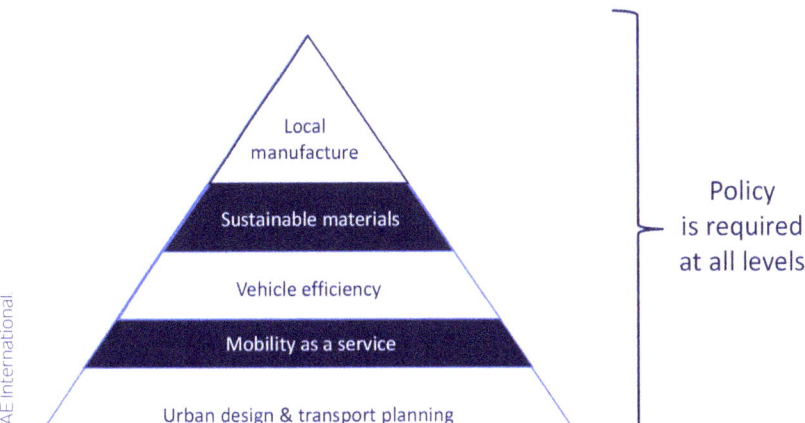

In essence, the proposed new framework comprises the following elements:

Rethinking Auto-Mobility (Chapter 7)

- Urban planning and transport infrastructure
- Automated MaaS
- Last-mile goods delivery

Rethinking Vehicle Design and Development (Chapter 8)

- Vehicle right-sizing
- Solar power
- The autonomous, connected, electric (ACE) vehicle platform

Rethinking Vehicle Materials and Manufacturing (Chapter 9)

- A new approach to materials
- Micro-factories
- Government and community policy

The interplay of these elements creates a comprehensive framework for achieving sustainable and affordable auto-mobility. By prioritizing urban planning that supports active mobility, integrating MaaS, designing appropriately sized vehicles, harnessing renewable energy, and fostering a circular economy, cities can enhance mobility for all its residents while tackling environmental challenges. To gather more local support for this initiative, vehicles can be designed, developed, and manufactured locally to create high-quality employment and further improve sustainability. This holistic approach not only improves transportation efficiency but also contributes to healthier, more vibrant civic communities.

While the argument for reframing future mobility around these principles is logical and has potential benefits, it will face significant challenges in implementation. Consumers may be wary of vehicle performance limitations and reluctant to turn away from personal ownership, particularly in car-centric cultures. Local, customized production may have high initial costs and fail to achieve the economies of scale to lower vehicle costs sufficiently. Investment in integration with existing public transport may be difficult to achieve, particularly if vehicles have to be electrically recharged. These challenges will be discussed in more detail in later chapters, with potential solutions, such as fleet ownership and goods delivery.

The success of this new approach will depend on a mix of technological advancements, policy support, consumer acceptance, and the ability to overcome economic and logistical hurdles. A balanced approach that gradually incorporates these ideas while addressing public or corporate resistance may be needed in some regions more than others as the approach can vary from one part of the world to another. For example, in developed economies, autonomy can be leveraged to take friction out of the rental process while affordable solutions for last-mile operation in emerging economies may harness solar power and low-cost labor instead. However, as autonomy, connectivity, electrification, renewable energy and materials become mainstream, it is likely that similar solutions will be enabled and exist all over the world to provide more sustainable, affordable auto-mobility solutions.

OEMs have an opportunity to renew Henry Ford's promise for the twenty-first century by integrating automobiles with public transport in new, creative ways (Chapter 7), developing new types of vehicles and applying vehicle know-how to create new transport solutions (Chapter 8), and partnering with cities around the world to locally make and use new vehicles while tackling material waste (Chapter 9).

References

6.1. BloombergNEF, "The World Hit 'Peak' Gas Powered Vehicle Sales in 2017," January 30, 2024, https://www.bloomberg.com/news/articles/2024-01-30/world-hit-peak-gas-powered-vehicles-as-evs-gain-market-share.

6.2. Lavelle, M. and Gearino, D., "Competing Visions for U.S. Auto Industry Clash in Presidential Election, with the EV Future Pressing at the Border," Inside Climate News, October 30, 2024, https://insideclimatenews.org/news/30102024/us-auto-industry-ev-future-clash-at-the-border/.

6.3. "Huawei and SAIC Announce Joint Electric Car Brand Shangjie", Max McDee, 17th April, 2025, Arena EV, https://www.arenaev.com/shangjie_is_new_joint_electric_car_brand_from_huawei_and_saic-news-4625.php.

6.4. Wayland, M., "GM Can Regain Market Share in China after Hitting 20-Year Low, Executive Says," CNBC, May 9, 2024, https://www.cnbc.com/2024/05/09/gm-can-regain-market-share-in-china-after-20-year-low-exec-says.html.

6.5. Wu, S., Leussink, D., and Steitz, C., "Volkswagen Aims to Keep China Market Share Stable as Price War Rages," Reuters, April 24, 2024, https://www.reuters.com/business/autos-transportation/volkswagen-aims-maintain-15-market-share-china-2030-country-chief-says-2024-04-24/#:~:text=Volkswagen%20ceded%20its%20title%20of,as%20combustion%2Dengine%20sales%20declined.

6.6. Khatib, M., "US Investments in Electric Vehicle Manufacturing 2024," Atlas Public Policy, August 2024, https://www.atlasevhub.com/wp-content/uploads/2024/08/2024.08_Atlas-EV-Investment-Brief.pdf.

6.7. "Current Ambitions: Three Asian Countries vie to be the Next Electric-Vehicle Superpower," Economist, March 22–28, 2025, 62-63.

6.8. Kang, L., "Tesla Sells 75,805 Cars in China in Dec, Exports 18,334 from Shanghai Plant," CNEV Post, January 9, 2024, https://cnevpost.com/2024/01/09/tesla-sells-75805-cars-china-dec/#:~:text=Tesla's%20Shanghai%20factory%20exported%20344%2C078,data%20compiled%20by%20CnEVPost%20showed.

6.9. Ireson, N., "The Cheapest New Cars of 2024," Kelley Blue Book, December 18, 2023, https://www.kbb.com/best-cars/cheapest-new-cars/.

Chapter 07

Rethinking Auto-Mobility

7.1.
Urban Design for Mobility—the Ugly, the Bad, and the Good

Urban planners and transport officials must face the constant challenge of balancing economic development, accessibility for all, and environmental sustainability. During the last 100 years or so, as automobile usage has risen, it can be argued that several mistakes have been made, particularly in the car-centric US. These include the prioritization of space for vehicles to the detriment of pedestrian-friendly public spaces and the growth in urban sprawl that increases reliance on cars. On average, each of the approximately 300 million cars in the US requires three or four parking spaces, and the accumulated land set aside for parking space exceeds 5000 square miles in the US (and this does not even include the space for maneuvering into parking spaces inside parking lots, which may triple the area allocated to parking). Along with this excessive accommodation of the car, there is a simultaneous neglect of public transport, which not only includes limited funding but also a failure to integrate different modes of transportation effectively and to focus on a radial network (suburb to city center) rather than a comprehensive grid.

One of the root causes of suboptimal urban mobility has been zoning policy, whereby a strict separation of residential, commercial, and industrial areas has been encouraged, even fostering development in areas that are poorly served by public transit. The effects of zoning policy have led to increased travel distances and a need to use either a car or public transport because walking and cycling are not practical alternatives. Unfortunately, the resulting low-population-density zoning in suburbs makes efficient public transport more difficult and increases car dependency, which is a challenge for people with low income or disabilities. Because there is no attractive alternative that seamlessly connects public transport, cycling, and walking, transportation systems have often failed to serve disadvantaged communities, leading to disparities in economic opportunities and access to services. It has also made the city less safe for pedestrians and cyclists.

Prioritizing the road infrastructure over public transport led to increased car dependency and traffic congestion. A congestion charge has notably been enacted in Singapore, London, and several other cities worldwide to improve or simply maintain traffic speeds and to use the funds that it raises to improve public transport, but such efforts to combat congestion, even in the most densely populated US city, New York, have until recently failed in North America although this may be changing. Priority to roads and cars has been exacerbated by the building of an oversized road infrastructure that was often based on inflated traffic projections and in underestimating the potential for any shift to public and active transport.

Highways have often been constructed to go right through urban centers such as Los Angeles, which is known for its extensive freeway system that runs through and around the urban core, and Atlanta has 14 lanes of traffic that go through the downtown area. These policies have often destroyed existing communities. About one-quarter of the space in many downtown areas is devoted to parking for cars [7.1] (if road space is added, then this rises to more than 30%)! This allocation of space at the expense of parks, plazas, and pedestrian-friendly areas has, arguably, reduced the quality of life and limits the opportunity for social interaction (**Figure 7.1**).

Figure 7.1 Wide highways and large parking lots are common features of a car-centric culture, evident in many US cities.

City policy has tended to focus on short-term solutions rather than sustainable, long-term urban development. Funding has often supported a transport infrastructure that is difficult to adapt to changing needs and technologies. This creates an overreliance on single modes of transportation instead of multimodal solutions.

There has also been insufficient planning for the rise of freight deliveries, driven by e-commerce, and many cities are failing to plan for technological change, such as AVs, that could significantly reduce the need for parking space while putting extra pressure on the curbside for pickup and drop-off access. How will we manage the curbside when self-driving buses, robotaxis, and delivery vehicles are also vying for the space used today by personal cars and micromobility?

The neglect of active transportation (lack of walking and bicycle paths) has made it difficult to integrate these healthy, active modes with public transport systems. Parks and historic neighborhoods have been destroyed and eliminated walkable urban life. Low-income communities have often been displaced to accommodate urban renewal, and the uneven distribution of transport services has often neglected poorer neighborhoods and people with needs, such as the elderly and the disabled. On top of this, environmental concerns (noise, air quality, greenhouse gas emissions) have often been ignored or treated as an afterthought.

Recognizing these past mistakes is crucial for current and future urban planners and transport designers. Many progressive cities in North America are now working to correct these issues by implementing mixed-use zoning, promoting transit-oriented development (TOD), investing in public transport and active transport infrastructure, and focusing on creating more livable, sustainable urban environments. In 2020, Vancouver, for example, decided to reallocate 11% of road space to support public transport, pedestrianization, and micromobility [7.2]. In many respects, progressive cities are looking to Europe as a model for developed, densely populated cities that have a high quality of life enabled by effective transport solutions. Rapidly developing cities around the world, particularly in eastern Asia, are also looking to learn from US cities what **not** to do, as a cautionary tale.

It must be recognized that urban planning, not smart city technologies, provides the foundation for all mobility solutions and it involves designing cities to minimize travel needs in the first place. The superblocks in Barcelona are a good example of residential areas that are designed around pedestrians and cyclists, not cars. Superblocks are created by closing off all the inner streets to cars in a 3×3 block grid, diverting the traffic to the outer streets (**Figure 7.2**).

Figure 7.2 Barcelona's superblocks [7.3].

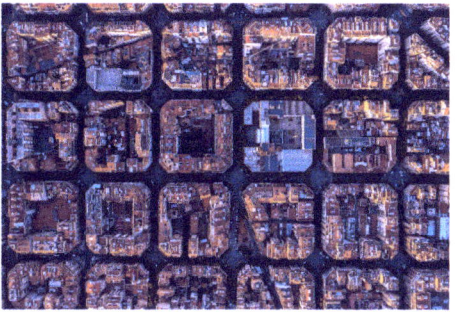

When the inner streets have become "pacified," they can become a communal walking and meeting space with opportunities for greening and shading. In practice, limited vehicle access is allowed, primarily for accessing homes and for emergency vehicles, but there are restrictions on vehicle movement (one-way traffic), speed (below 10 kph or 6 mph), and parking. People with disabilities or those needing to deliver heavy items can still reach their homes by car, but the superblock design aims to balance the needs of all residents while promoting a more livable urban environment. In Chapter 8, Section 8.5 and Chapter 9, Section 9.5, consideration will be given to extending this concept further to "car-free" city centers.

Many European cities are walkable and can be considered "15-minute cities," meaning that most amenities, such as schools, hospitals, stores, and restaurants are within a 15-minute walk (one mile distance) for nearly everyone. To encourage active modes, such as walking and cycling, there needs to be investment in infrastructure such as dedicated bike lanes, wide sidewalks, pedestrian bridges or underpasses, and traffic calming measures to make it safer and more appealing. Safe pedestrian crossings, attractive streetscapes, protected bike lanes, bike parking facilities, and bike-sharing programs can also help encourage people to adopt active modes instead of always defaulting to car usage. A good example of this people-centric design, where the focus is on a human scale and experience, prioritizing pedestrians and cyclists, is Copenhagen, which has, perhaps, the most extensive bicycle infrastructure in the world (**Figure 7.3**). As cyclists dominate the roads and feel safe in numbers, driver behavior also changes to become more cautious, which reinforces safety for cyclists even more.

Figure 7.3 Bicycle parking in Copenhagen.

Mixed-use zoning in urban areas combines residential, commercial, and recreational spaces in the same areas and aims to reduce the need for long-distance travel for daily activities. Amsterdam's compact, mixed-use neighborhoods are a good example of how to integrate diverse functions to reduce travel needs and create vibrant neighborhoods with good connections between them via walkable streets, bike lanes, and public transport.

However, for longer journeys that cannot easily be accomplished by walking or cycling, there needs to be public transport. TOD concentrates housing, commercial areas, and public facilities around public transport hubs, reducing car dependency and promoting the use of public transport [7.4]. TOD encourages higher population and employment densities in areas close to transit stations and supports the viability of public transport by ensuring a steady demand for services. Zoning laws and building codes can be modified to allow for taller buildings that are mixed-use so that people are living above the retail or office space, increasing land efficiency and creating a more vibrant street-level environment. By attracting business in transit-adjacent areas, property values are often higher. This means that to ensure equitable development, TOD strategies often include provisions for affordable housing, which make it possible for people of different income levels to live near public transport stations (**Figure 7.4**).

Figure 7.4 TOD.

TOD can occur in cities of all sizes, including megacities, like Tokyo, where there is a lot of development concentrated around major train station hubs. Cities can take other actions to reduce car dependency or to make driving less attractive. As an example, strategically managing parking through reduced parking minimums, shared parking structures, or even car-free zones can discourage excessive car use. These policies do not require any new technology to be developed, but it should be noted that successful urban design practices may not always be transferrable because they need to consider local factors such as climate, culture, and existing infrastructure and should be adapted accordingly.

Not all the public transport innovation has occurred in the developed world. Bus rapid transit (BRT) is a cost-effective alternative to underground subway systems and was pioneered by Jaime Lerner, a noted urbanist and Mayor of Curitiba (Brazil), where he introduced it in 1974. Bogotá adopted BRT in 2000 and currently has the largest BRT system in the world with its TransMilenio BRT system (**Figure 7.5**). It handles 2.4 million passengers a day along its network of 143 stations and 71 miles [7.5].

Figure 7.5 Bogotá TransMilenio BRT.

With dedicated bus lanes that are separate from the other traffic, elevated stations with real-time information displays, and fares that are collected prior to boarding to reduce delays, BRT mimics a subway system but with far less capital cost. The system has been credited with reducing travel times, air pollution, and traffic congestion.

Although public transport should provide more sustainable and space-efficient mobility solutions than personally owned vehicles in densely populated urban areas, they routinely suffer from several shortcomings that impact their overall convenience and usability. These include a low frequency of service outside peak commuting hours, limited route coverage in the suburbs, long wait times for transferring between different routes or modes (e.g., bus to rail), and a lack of seamless first-/last-mile connections between public transport stations and a person's actual start or end point, which often requires walking, cycling, using automobiles, or other transport modes. As the population ages and becomes increasingly obese, more attention will need to be given to ensuring that all community members have access to mobility solutions. Equitable access to transport options is essential for fostering inclusivity in urban environments. Ensuring all fixed-route public transport systems meet ADA requirements, and implementing paratransit services and shared-ride programs is a start, as is offering the elderly and the disabled reduced fares. Sidewalks, crosswalks, and public spaces also need to be accessible and safe for individuals with mobility challenges.

Although it does not provide door-to-door connectivity, Tokyo has perhaps the most extensive public transport system in the world. In the central 23 wards of Tokyo, where nearly ten million people live, the average walking distance to a train or subway station is typically under half a mile. This means that in dense urban areas like Shinjuku, Shibuya, or Minato, one can almost always see a station entrance from anywhere! In the broader Tokyo metropolitan area, which includes surrounding prefectures like Saitama, Chiba, and Kanagawa, the maximum walking distance might extend to a mile in less densely populated suburban regions. However, even in these areas, public transport coverage remains exceptional. The Tokyo Metro and Toei Subway systems alone have 285 stations across 13 lines. When the JR East train lines, private railway networks like Tokyu and Keio, and regional train systems are included the total number of train stations in the greater Tokyo area exceeds 900. Then there is an extensive bus system, including Toei Bus, that can connect areas not directly served by the train (**Figure 7.6**).

Figure 7.6 Tokyo's world-class rail system.

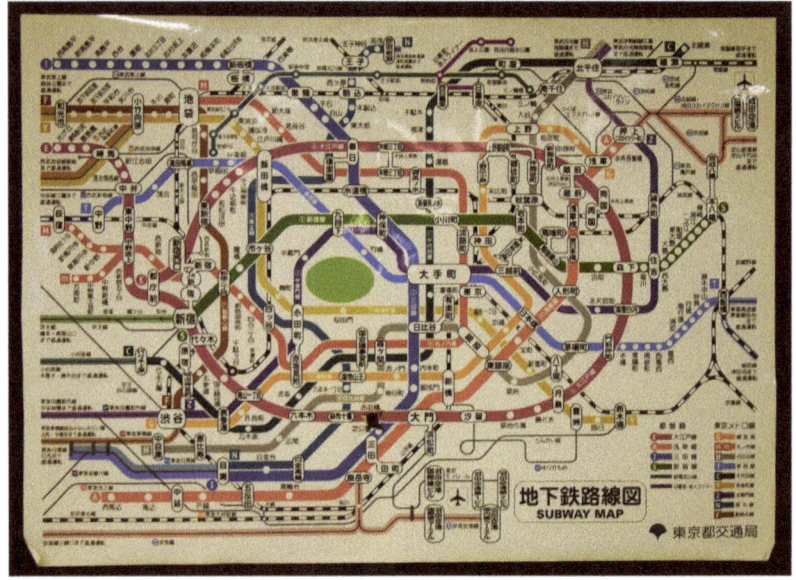

Interestingly, the Japanese government and transit authorities have deliberately planned the areas around train stations (TOD, in other words) so that residential and commercial districts are strategically built around train stations, further reducing walking distances and making public transport incredibly convenient. It has integrated the ticketing process across different modes of transport, including trains, subways, and buses as well as in other parts of Japan. Major bus stops are often located near train and subway stations, making it easier to transfer between modes, and transit apps and information boards often provide combined information for trains and buses. This coordination between buses and trains means that when trains stop running late at night, buses will often serve as an alternative and they often provide feeder services to train stations in suburban areas. This level of integration occurs because efforts have been made to create a relatively seamless experience for passengers, despite different companies operating various parts of the system. In summary, this form of multi-modal integration coordinates different public transport modes (e.g., buses, trains, trams) to create a seamless network with coordinated schedules, ticketing, and information systems. It tends to be operated by city public authorities or through public–private partnerships, and the primary goal is to increase ridership and revenue by making public transport more convenient and efficient.

Emerging technologies can play a big role in improving the public transport experience and reducing car dependency. For many people, access to free Wi-Fi while riding public transport leads to a better use of time than driving a car and searching for parking. Emulating ride-hailing apps by providing bus and train riders with real-time information on location and estimated arrival time can allow people to plan their time accordingly even if they do have to wait. London's use of real-time data to manage traffic and public transport is a good example of this.

In the near term, some cities are promoting themselves as smart because they are using sensors, connectivity, and big data analytics to monitor and optimize traffic flow, parking, and public transport in real time. Intelligent traffic systems can allow traffic lights to respond to current traffic conditions, and dynamic road pricing can be used to manage travel demand in real time, and was pioneered in Singapore in 1975 with the current electronic version beginning in 1998 (**Figure 7.7**).

Figure 7.7 Singapore's Electronic Road Pricing relies on both infrastructure gantries and in-vehicle units (IU) for dynamic payments.

These solutions not only improve traffic flow today, but with machine learning, they can be used to optimize transport networks, improve user experience, provide a better understanding of mobility patterns, and help cities design better services and infrastructure for the future that can serve actual demand. In the long term, AVs could lead to more efficient traffic flow, improved safety, and better access for the underserved (or not, as discussed in Chapter 5, Section 5.6.2), but only if they are properly leveraged and implemented, as described in Sections 7.2 and 7.4.

Cities can also try to affect behavioral change. Although this is not easy, it could make a big difference because public engagement and educational campaigns can promote the benefits of using active modes, public transport, and shared mobility services, and foster a culture of civic virtue and of sustainability. Complementing this emphasis on health and sustainability is an attention to greening the infrastructure. For example, cities can encourage renewable sources of energy, such as solar and wind, to charge the EVs used by their urban fleets such as buses, trains, as well as shared e-bikes and scooters. They can also incorporate green spaces and sustainable stormwater management systems, as has been done in Portland (Oregon) with its Green Streets program that combines stormwater management with a pedestrian-friendly design.

7.2.
Autonomy and Public Transport

Although electrification can support public transport by reducing energy and maintenance costs and by supporting clean air and climate change initiatives, it will probably not have as much transformational impact on the operational side as autonomy. Autonomous buses and trains have the potential to offer a cheaper, more frequent, flexible, and reliable service, attracting more people to use public transport. Compared to robotaxis, robobuses can reduce congestion, and autonomous operation should be easier to achieve because the cost of the autonomous stack hardware is easier to absorb in the vehicle cost, and also because the vehicles operate on a fixed route that can be more easily repeated and machine-learned. AVs can analyze real-time traffic conditions, passenger demand, and vehicle locations to optimize routes and schedules dynamically. This should lead to reduced travel times, less idling and lower energy consumption, and fewer empty vehicles.

It can achieve this because it eliminates the need for a human driver and the costs associated with this, which amount to roughly 40% of the total operating costs for a bus [7.6].

In addition, by optimizing the schedule and routes, operating for longer hours without breaks to increase vehicle utilization and asset efficiency, and by reducing accidents and downtime, autonomous public transport could become more cost-effective, potentially leading to lower fares for passengers while providing a higher quality of service. It could, for example, increase the night-time schedule frequency or even enable on-demand operation, reducing inefficiencies in public transport operations when it is heavily underutilized.

Smaller AVs (roboshuttles, robotaxis, and/or robopods) can provide microtransit services, filling gaps in traditional public transport networks and serving areas with lower population density. They can complement the public transport system by providing transportation for people with disabilities and those without a driver's license. Smaller AVs can be used for less-traveled routes and can adapt to fluctuating passenger demands. They have lower passenger capacities, making them suitable for urban areas where demand varies significantly throughout the day. This adaptability allows for a more tailored service compared to larger buses, which may run on fixed schedules regardless of actual passenger needs. These smaller AVs could facilitate trips in low-income communities that were previously impossible or impractical by providing affordable, last-mile transport for people who are unable to drive. This could dramatically improve mobility for the poor, the elderly, and the disabled.

With wireless communications between them, several smaller AVs could platoon together and even transport more people than a bus, or more goods than a large truck, but with much greater flexibility and efficiency. Platooning allows for the dynamic adjustment of vehicle numbers based on real-time demand, so small AVs can easily split and reconnect, providing a flexible response to varying passenger loads throughout the day. This capability is particularly beneficial in urban environments where passenger numbers can fluctuate dramatically. Moreover, platooning can reduce aerodynamic drag and lower energy consumption for each vehicle (**Figure 7.8**).

Figure 7.8 Platooning of smaller vehicles could lead to more efficient and cost-effective public transport than large capacity buses.

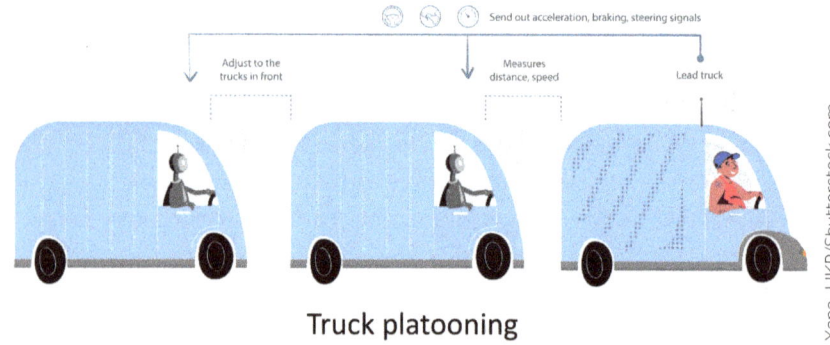

Truck platooning

When comparing how these smaller AV fleets can serve urban mobility needs and provide last-mile connectivity with public transport, most of the attention centers on the robotaxi, but it is far from clear if this is the best solution because it may, in fact, compete with public transport as much as it serves; robotaxi operation, after all, is being driven by commercial interests! Like ride-hailing, much of the travel demand for robotaxis is expected to be in the already-congested city center, rather than in underserved areas that lack access to public transport. Collaboration between city planners, public transport authorities, and robotaxi companies will be necessary if robotaxis are going to complement and integrate with existing public transport systems. For example, the city could enact a congestion tax or fee on robotaxi rides if they operate at times that contribute to congestion, and they could use the proceeds to support subsidies for robotaxi rides in underserved areas for low-income residents. This would allow these people to benefit from the new transportation option while being less costly to the city than existing microtransit options. The impact of cheaper robotaxi rides on congestion and their benefit to underserved populations ultimately depends on how they are integrated into existing transport systems, but with thoughtful regulations, city planners can try to make robotaxis a force for good, reducing congestion and providing a much-needed mobility option for all. Moreover, in the long run, sharing anonymized data on travel patterns between robotaxi companies and public transport authorities could help to inform how to optimize the overall transportation network and improve the future transport service in underserved areas.

When considering solutions for underserved areas, other AV options that should be considered include roboshuttles and robopods (**Figure 7.9**). Roboshuttles are autonomous shuttle buses that follow predefined routes, which makes the autonomy technical challenge easier to solve. Roboshuttle developers, such as China's WeRide and US-based May Mobility, are currently operating in Singapore and Detroit. Europe, where French-based Navya and EasyMile are developing and operating roboshuttles, may prove to be the most attractive market because of the regulatory environment and strong public transportation focus (**Table 7.1**). Renault, for example, is partnering with WeRide in Valence (France) to connect its TGV train station with the surrounding business park along a two mile route [7.7].

Figure 7.9 Autonomous shuttle operating in Helsinki.

Table 7.1 Robo-X: A comparison of robotaxis, roboshuttles, and robopods.

	Robotaxi	Roboshuttle	Robopod
Operational demand	Most roads	Fixed routes and schedules connecting to public transport stations	On-demand and within a few miles radius of public transport stations
Typical vehicle size (number of passengers)	5–7	10–15	2–4
Scalability	Passenger vehicle economies of scale	Lower-volume vehicles	May share economies of scale with microcars or small cars
Cost	High development costs because of technical complexity	Lower development costs	Lower development costs
Safety	More likely to be involved in traffic and safety incidents	Operates at lower speeds and in quasi-dedicated lanes	Operates at lower speeds, may need to meet Federal Motor Vehicle Safety Standards (FMVSS)
Public acceptance	May take jobs away from ride-hailing and taxis	Improves public transport and accessibility	Improves public transport and accessibility
Current status	Extensive trials in a few locations	Limited trials in a few locations	Concept

© SAE International.

Unlike commercially operated robotaxis, roboshuttles are more likely to be seen as part of a public transport system and can provide "close" to door-to-door transport service cost-effectively because the ride is shared with a relatively large number of passengers compared with a robotaxi. However, it tends to operate on a fixed schedule, and a lack of funding to move beyond limited testing has reduced the number of roboshuttle developers in recent years [7.7].

Robopods, meanwhile, do not really exist yet but are autonomous microEVs that can, potentially, be based at or near existing public transport stations and provide first-mile/last-mile transportation solutions on demand, helping people in underserved areas reach public transport stations. Instead of adding to congestion in city centers, as robotaxis are likely to do, robopods could provide last-mile connectivity in the suburbs and encourage public transport usage. In Milton Keynes (UK), four-passenger autonomous pods are being trialed that travel approximately a mile from the train station to the shopping center along a dedicated pathway that is kept separate from car traffic. Personal rapid transit (PRT), where pods operate on guideways that are separated from the road, is another approach that is being used at London's Heathrow Airport Terminal 5 to connect it with the parking lot about two miles away (**Figure 7.10**). Although PRT can leverage the same technologies that are being developed for AVs and typically operate in a simpler environment where there are no pedestrians, cyclists or other vehicles, it does require a dedicated guideway or path infrastructure that makes scaling far more expensive to achieve. The vision of robopods, however, is for them to operate on normal roads, or perhaps bus and bike lanes, leveraging all the advances in robotaxi development. Unlike roboshuttles, robopods are still largely a concept but one that may be compelling as part of a MaaS solution as they provide the benefits of carsharing with autonomous operation so that they can be picked up and dropped off very conveniently and without requiring a driver's license.

Figure 7.10 London Heathrow Airport's Ultra Pod PRT system connects Terminal 5 to the business passenger parking lot, providing a more timely and predictable service than regular shuttle buses.

As these systems evolve, they have the potential to significantly reshape personal mobility, especially in areas where the need for flexible, on-demand transportation is highest. The integration of autonomy and on-demand mobility solutions can directly address many of the weaknesses in public transport. Higher vehicle occupancy rates and more efficient use of road space could be achieved by using fleets of small AVs to dynamically route and pool riders who are headed in the same direction, rather than running mostly empty large buses on fixed routes. The frequency of service could be substantially increased by dispatching AVs as needed, based on real-time demand patterns rather than operating on rigid, scheduled timelines that are often set months in advance.

The seamless integration between traditional high-capacity public transport (rail, bus, and BRT) and on-demand smaller AVs can create a cohesive, multimodal mobility solution that plays to the strengths of each component (**Figure 7.11**). For example, the small footprint and tight turning radius of compact AVs (robopods) could enable them to effectively serve as first-/last-mile feeders into public transport stations without contributing as much to road congestion because they would operate mostly in the less densely populated suburban areas. Their autonomous capabilities would allow for intelligent dispatch and routing to synchronize their arrival and departure with the scheduled bus and train service. And their zero-emission operation allows them to freely access pedestrianized city centers, environmental zones, and even indoor locations that may be restricted for conventional ICEVs.

Figure 7.11 Public transport integration facilitated by vehicle right-sizing, autonomy, and electrification.

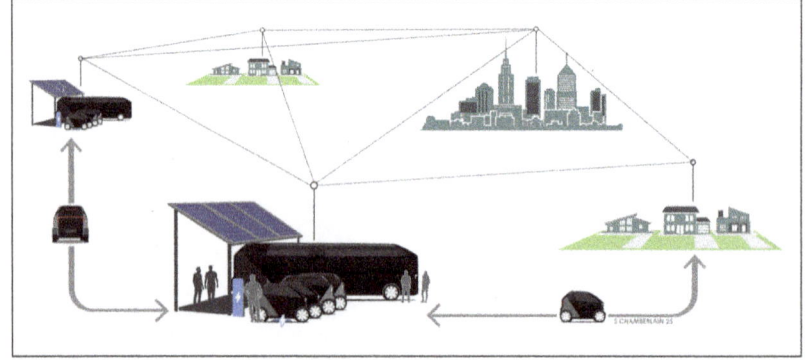

Cities wanting to safeguard affordable access should seize on this new technology to rethink land use patterns and promote denser, more walkable neighborhoods. AVs can generate vast amounts of data that can be used to help transport and urban planners make better decisions about future infrastructure, services, and resource needs. By embracing these technologies and learning from past mistakes, urban planners and transport officials can create more sustainable, equitable, and user-friendly public transport systems that meet the evolving needs of their customers or "voters." This will increase the utility of MaaS and create the conditions to significantly reduce household dependency on vehicle ownership.

7.3. MaaS

As discussed in Chapter 3, Section 3.7, MaaS usually involves a subscription or pay-as-you-go model for accessing the various mobility services. The aim is to provide a seamless journey planning, booking, and payment process across different modes of transport, heavily reliant on the smartphone. Multimodal passes allow users to access various modes of transportation within a city or region, including the common forms of public transport (e.g., buses, trains, and subways), and with possible access to shared mobility services like car-sharing or bike-sharing. Connectivity and the smartphone have been critical to making multimodal passes feasible because they allow users to plan multimodal journeys more effectively in advance, which provides a more seamless transportation experience that can go "door-to-door" or close to it within a defined transport area. Compared to owning a car, it is promoted as being less expensive and more convenient, especially when parking is factored in, better for the environment and a healthier option. Cities in Northern Europe and developed parts of Asia have led the way in integrating different modes of transport.

MaaS builds on multimodal integration and is a relatively new concept in transportation. Most MaaS platforms are not profitable at the moment and are still in the experimental or growth phase, with a focus on increasing the user base rather than profitability, and many have struggled with profitability despite impressive user growth. MaaS revenues tend to come from user subscription fees, commissions from transport providers, data monetization (with privacy considerations), and partnerships with cities or businesses. The market is expected to grow significantly in the coming years, and as systems reach economies of scale, the goal is for profitability to improve. In the meantime, profitability is challenged by the high development and maintenance costs for the MaaS platform itself, managing many partnerships with the various transport providers and, sometimes, navigating regulatory complexities across different regions. Of course, as a new concept, it must also promote itself to users and change their behavior to some extent. The path to becoming profitable probably involves city government support and subsidies because MaaS has real potential to increase public transport ridership, and achieving a critical mass of users is in both their interests. While not currently monetizable, MaaS adds value to the city by reducing congestion and emissions, improving overall transportation efficiency and accessibility, and generating valuable data for urban planning. In summary, MaaS is still in a developmental stage, business models are continually being refined and profitability remains a challenge for most platforms, but it shows promise and is likely to grow in importance over time.

Before discussing what it will take to make MaaS an extremely convenient, practical, and affordable alternative to owning a car, it is worth remembering that it is possible today to hire a taxi or rent a car for those occasions when public transport and active modes are not attractive, but this practice has not been commonly used except in densely populated cities or when traveling by air to another city. For everyone else who has the means to afford a car, the benefits of vehicle ownership typically outweigh the disadvantages that come with it, such as the purchase or lease cost, insurance, parking, refueling, and maintenance, as illustrated in **Figure 6.8**. Owning a car has been associated with a "rite of passage" and has given a sense of freedom to "go anywhere, anytime, with anyone." Personal vehicles enable door-to-door travel for several people and their cargo, often providing the fastest travel time barring significant traffic delays. A corollary of this is that the vehicle is overengineered for typical everyday use, as was covered in Chapter 2, Section 2.1.2. The other benefit that automobiles provide is a personal space to feel "cocooned." Having a quiet space that is free from fellow riders and being able to personalize climate and infotainment settings while having secure storage for possessions may be some other reasons that encourage customers to purchase automobiles. Having said that, many young consumers, in particular, are beginning to view personal vehicles as mobility "appliances" rather than as status symbols. For these people, driving is a distraction from being connected, and the vehicle's infotainment system is becoming a more important purchase consideration than exterior styling or branding. This could mean that a growing number of vehicle owners may be open to either sharing their vehicle peer-to-peer or using a vehicle subscription service solution themselves. After all, owning a high-fidelity audio system and buying records used to be important for many consumers, whereas now streamed music is purchased and listened to on smartphones.

What if someone could choose any vehicle they want or need at any given time, based on that particular situation, and have it available immediately on demand? As mentioned in Chapter 2, Section 2.1, privately owned vehicles are typically parked 23 hours a day, and this suggests that there should be an opportunity, especially with today's instantaneous connectivity and location sharing, to rent one's vehicle to another person who needs to borrow it for a certain period of time, such as overnight (when the owner sleeps), during the day (when the owner might be at work), or for a week (when the owner is away on vacation). Vehicles are effectively idle assets that depreciate the moment they are purchased, but with connectivity, they can become a source of income. With the sheer number of nearby vehicles that are not being used at any moment (parked), it may be possible to create a marketplace in the future that links vehicle owners to vehicle users and handles insurance, cleaning, refueling, repair, etc. to mutual satisfaction. In this scenario, the consumer can forego the disadvantages of car ownership while having access to a wider range of vehicle options tailored to their specific needs so that they are always driving the best vehicle for the task.

Or someone could choose to own an affordable, efficient compact urban EV that serves most of their travel needs within the metropolitan area, but still have on-demand access, via a mobility service provider, to a full-size pickup truck, van, or SUV as needed for utility, travel, and recreational purposes. If a solution can be found for this challenge, it will make MaaS a truly viable alternative to owning a car because it would allow on-demand access to all types of vehicles (automobiles and micromobility solutions) as well as seamless integration with public transport, with real-time pricing and information on the best mode of travel, based on speed, cost, and so on.

The intrinsic appeal of this idea means it has been studied for decades, but it has been difficult to implement, for a variety of reasons. In the not-too-distant future, autonomous operation should solve one of the major challenges, vehicle pickup and drop-off, and also drive the vehicle to charging stations to recharge and then autonomously return to the fleet location, ready for reuse without requiring any human intervention in the charging process. This could perhaps be eventually extended to automated washing and cleaning as well (**Figure 7.12**).

Figure 7.12 The consumer, not the vehicle, is at the center of the new mobility model.

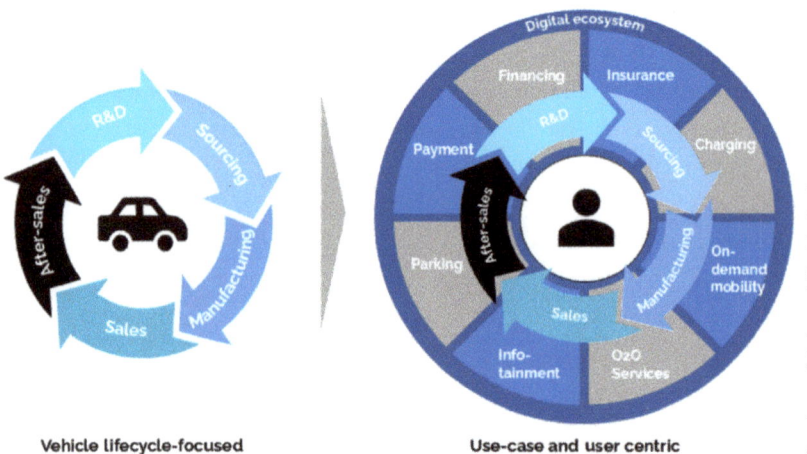

As with EVs, and potentially with AVs, the country that might lead the world in developing and implementing such a sweeping new vision of mobility, away from a vehicle ownership model and toward a vehicle accessibility model, is China. They possess Internet-savvy consumers and companies, such as Tencent, Alibaba and Baidu, that are motivated to focus on the digital experience even more than the driving experience. They have the example of a lucrative smartphone business model recently in mind, with significant updates and revenues available post-sale. Expanding this data-monetizable approach to mobility is a logical and compelling step and is tied to the development of SDVs. China also has a strong interventionist government that could further accelerate integration between personal mobility and more affordable public transport options.

As the population ages and becomes more obese, the need to provide mobility that does not rely on being able to drive becomes ever more important, and this could be the strongest argument for AVs. If automated MaaS is the most affordable and sustainable way to provide access to people, things, and places anywhere at any time, what needs to be done to remove the "friction" in the system that presently makes people buy overengineered and expensive vehicles and keeps them from adopting vehicle subscription services and MaaS? A potential answer lies in applying the same solutions that are currently having to be developed by robotaxi and autonomous truck companies in order for them to create a viable business and not just a viable technology.

7.4.
Removing "Friction" in a Vehicle Subscription Service

Creating a viable AV business involves far more than solving the autonomy challenge, even if that is the fundamental problem that must be solved. For example, software must also be developed that can efficiently route vehicles and assign them to ride requests across the fleet because a high proportion of "paid miles," rather than "empty miles," is essential for economic viability. It is critical to have the right balance of supply and demand and to implement dynamic or "surge" pricing to extract the most profit. Providing a large enough vehicle fleet while ensuring the vehicles are used efficiently to generate sufficient revenue is challenging, and designing systems that can scale from a small pilot program to large-scale operations is necessary to achieve a cost structure that allows for profitable operation at competitive prices. Balancing supply and demand with pricing structures that are attractive to consumers while still maintaining profitability is not easy.

Potentially integrating with existing transportation networks or other mobility services will require further coordination, and so will land purchases for vehicle storage overnight. The complex and evolving regulatory landscape for AVs must be navigated with local and regional government agencies, to obtain the necessary permits and licenses for operation. Issues concerning data management security and privacy must be addressed, and liability must be determined in partnership with insurance agencies.

In terms of dealing with other stakeholders, it is important to build trust and acceptance, which means there should be extensive promotional or educational campaigns concerning AV technology and the AV service. For the customer, there should be a userfriendly mobile app and a customer support hotline for handling issues or queries, such as pickup and drop-off locations.

The upfront costs of building out the required infrastructure and technology are also significant. From an operations standpoint, a fleet of AVs will require large support facilities for servicing, recharging, and cleaning them daily. Because AVs are likely to be EVs, the storage facilities may need to include EV charging stations or the AVs will need to access external EV charging infrastructure at the right time and place, potentially integrating with smart city infrastructure. Staging areas and intelligent parking solutions will be required to avoid congestion and to efficiently manage AV fleets. Remote monitoring of the AV fleet is required with potential for intervention, at least for the foreseeable future, to handle edge cases.

Many of the requirements mentioned above will also need to be in place to remove "friction" from a vehicle subscription system, in order to make it far more convenient for people to access vehicles without ownership. As with robotaxis, acquiring and maintaining a large fleet of vehicles for a subscription service is costly, ensuring vehicle availability and managing maintenance across different locations can be challenging, and efficient vehicle distribution and rotation are essential for meeting customer demands (**Figure 7.13**). As autonomy proves itself, the need to have specific pickup and drop-off locations can be relaxed because AVs can, in principle, be picked up and dropped off anywhere and then be automatically routed to pick up the next user, leveraging predictive algorithms for demand forecasting and vehicle placement.

Figure 7.13 Waymo's robotaxi fleet maintenance center in Phoenix, Arizona.

A partnership with a parking facility may be required, with on-site maintenance and recharging capabilities. The vehicles must be monitored in real-time for maintenance, security, and cleaning purposes and to support usage-based insurance models. Predictive maintenance can help with vehicle reliability, and the vehicle could be cleaned with an automated system or with a quick-turnaround professional cleaning service.

Optimized vehicles for subscription service use, having a lower cost per mile of operation, should be designed with interior materials that are easily cleaned and have standardized vehicle interfaces across all different vehicle models with easy-to-use controls. Peer pressure on social media, user ratings, and gamification can also encourage users to keep the vehicles clean and be responsible.

A vehicle subscription service will need a user-friendly digital platform and mobile app that makes booking a vehicle easy and provides real-time availability, pricing, and location tracking, on top of advanced identity verification, digital keys for keyless vehicle access and operation, and automated billing and integration with digital wallets. Blockchain technology can provide secure, transparent recordkeeping for transactions and vehicle history, to ensure data security and privacy are protected. However, it should be recognized that the data gathered does provide valuable insight into usage patterns, which can inform future vehicle design and service improvements. For this reason alone, some data sharing and compatibility between various MaaS partners and technologies is needed, and there should be integrated payment systems across MaaS because subscription service users are likely to be among the heavier users of MaaS.

Personalized preferences can be stored, such as the ability to preset vehicle preferences (e.g., seat position, climate, music), and a range of vehicle types should be available to suit different needs, with surge pricing if certain vehicle types are most popular at a certain time. There should be options for users with disabilities and multilingual support for the broad customer demographics that are common in urban centers, as well as 24/7 digital support and quick response for emergencies and issues.

Vehicle subscription companies have faced higher insurance costs because of the higher risk that comes with multiple drivers using the same vehicle but, near term, manually driven vehicles in a subscription service that only operate in urban areas may be geo-fenced and speed-controlled, should limit insurance risk. Developing appropriate insurance models that cover both the company and subscribers will be complex but could be easier for zero-emission AVs that operate only in a metropolitan area, especially when the subscription service supports city government policies concerning the environment, congestion, public transport usage, and accessibility.

In summary, several emerging technologies and approaches could help to make vehicle subscription services more feasible for reducing household car ownership. The autonomous driving capabilities developed for robotaxis could be applied to vehicle subscription fleets, allowing the vehicles to transport themselves to subscribers without them having to travel to pick them up. Robotaxis rely on advanced fleet management systems to coordinate the deployment, maintenance, and recharging/refueling of the vehicles, and these same fleet management capabilities could be adapted to handle the logistics of a vehicle subscription service, ensuring vehicles are available when and where subscribers need them. Connected vehicles can have features like remote start, keyless entry, and real-time monitoring to enhance convenience and security.

Some companies are sensing an opportunity to focus on managing fleets, maintenance, and insurance for robotaxis and autonomous trucks, and this expertise could be transferable to vehicle subscription services. Examples include existing subscription programs, such as Hertz My Car, Nissan Switch, and Care by Volvo. However, these programs are typically for monthly terms and are not on a pay-as-you-go basis. They are also targeted at the car buyer, offering vehicles to be used anywhere, whereas a MaaS vehicle subscription might focus more on urban vehicle usage with potentially reduced cost and reduced requirements for insurance, maintenance operations, and vehicle and autonomy performance.

Allowing private vehicles to join a robotaxi fleet, as Tesla has proposed, eliminates the need for a massive upfront investment in a dedicated fleet as well as the associated vehicle storage, maintenance, and charging facilities because the owners will be responsible. It also allows Tesla to rapidly scale its robotaxi service across different locations, and there could, ultimately, be a wider selection of vehicle types to choose from.

However, this model creates some new challenges. For example, privately owned vehicles operating as commercial robotaxis could make liability in case of accidents or damage trickier to determine, and private owners may have to bear some of the responsibility for ensuring their vehicles meet local regulations for robotaxi operation. Ensuring all participating privately owned vehicles have the latest software and hardware and that they have consistent quality and cleanliness will be more challenging than with a centrally managed fleet. Managing and protecting data from privately owned vehicles used in the robotaxi service presents even more privacy challenges for both the owner and the riders. From an operational perspective, balancing the availability of vehicles with when the owners may want to use them at peak demand times may be more complicated because Tesla will have less control than with a dedicated fleet but there may also be many more vehicles available to cope with the demand. Tesla is likely to take a percentage of the revenue generated by privately owned robotaxis, similar to ride-sharing platforms, generating income without the full burden of fleet ownership and management. If urban autonomy can be solved by Tesla or other companies, then it may be feasible for vehicle owners to provide their AVs as part of an overall MaaS subscription service, complementing bike and scooter sharing as well as public transport.

Finding the right balance between affordability for consumers and profitability for companies is a challenge, and developing appropriate insurance models that cover both the company and the subscribers in case of accidents or damage will be complex. However, despite significant challenges, it should be possible to apply the emerging technologies, services and business models being developed by AV developers into making vehicle subscription services in some form or another increasingly feasible, especially if constrained to urban areas and integrated with MaaS.

7.5.
Goods Transport

The commercial transport and logistics sector, particularly the realm of urban and last-mile goods delivery, presents one of the most promising and fertile opportunities for the large-scale adoption of both EVs and AVs in the near future. US commercial vehicle market sales are projected to grow at a compounded annual rate of 10% to 2030 [7.8]. Powerful economic forces are driving this trend, including the need for faster and more efficient delivery times to meet rapidly evolving consumer expectations in the e-commerce age, persistent labor shortages affecting transportation and logistics companies, and the inherent suitability of EVs and AVs for operating on predictable, repetitive delivery routes within a defined geographic area where labor costs can be high relative to fuel and maintenance expenses.

Promising applications for autonomous operation in trucks include L2-/L3-assisted autonomy to improve safety and fuel economy, favored in China where labor costs are low, L4 autonomy on hub-to-hub routes in the US along highway corridors connecting major cities with reasonable weather all year round (e.g., the southern belt between Los Angeles and Atlanta), and L4 autonomy in cordoned-off private areas such as freight-handling ports (**Figure 7.14**).

Figure 7.14 Typical autonomous truck sensor suite and interior layout.

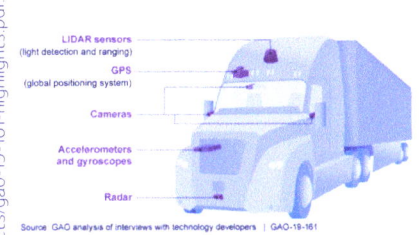

Another attractive opportunity for autonomous operation in goods movement is last-mile delivery. This refers to the final step in the logistics process, where goods are transported from a distribution center to the end consumer. This segment is crucial for both developed and developing countries, especially as e-commerce continues to grow. However, it poses significant challenges, including environmental impact, cost efficiency, and traffic growth.

Increased delivery volumes, especially from e-commerce growth, are leading to more vehicles on the road and generating more emissions. This is even the case if EVs are used because the energy may not all be from renewable sources. Last-mile delivery accounts for roughly half of the total shipping cost, and the average last-mile delivery cost was estimated at $10.10 per delivery in 2018 [7.9], eroding profits because of increasing costs and customer expectations for fast, free delivery. Driver wages are a significant component of last-mile delivery costs and retaining delivery personnel can be challenging. The demand for faster deliveries (same-day and next-day, for example) puts pressure on logistics operations and creates a need for flexible delivery options and precise delivery windows, while unsuccessful delivery attempts and managing returns add complexity and costs to the last-mile process.

Urban congestion makes it difficult to efficiently navigate delivery routes, and a lack of adequate parking or unloading zones in cities creates challenges for delivery vehicles. Moreover, there is limited capacity in the existing logistics networks to handle more delivery volume. Dynamic routing software and real-time tracking for better communication with customers are two solutions that can optimize operations, but there is a cost to implementing and maintaining these technologies. Companies are exploring various strategies such as offering more flexible delivery options and bundling deliveries, as well as adopting smaller vehicles, such as bicycles, cargo bikes, electric scooters, EVs, and drones for deliveries in congested areas and with less pollution (**Figure 7.15**).

Figure 7.15 Various last-mile autonomous delivery vehicles.

© General Motors.

Establishing small distribution centers closer to high-demand areas and collaborating with local authorities to create designated parking zones for delivery vehicles and advanced parking reservation systems may be viable approaches to providing timely delivery with minimal disruption to traffic and the curb. Internet of Things (IoT)-enabled smart lockers can allow customers to pick up packages at convenient locations, reducing the need for home deliveries, and crowdsourced delivery options are being trialed as well. In the long run, the city also needs to consider how to integrate delivery considerations into future urban planning requirements.

To alleviate road congestion during peak travel hours and the associated environmental impacts, some cities are looking at how to facilitate and incentivize night-time delivery options for businesses. Stockholm began an off-peak delivery pilot project in 2014 to better organize urban freight transport and reduce traffic congestion, while New York City has been researching the benefits of off-peak deliveries by implementing programs to encourage businesses to shift deliveries to overnight hours to reduce daytime truck traffic. Chicago, Sao Paulo, Paris, Stockholm, and several other European cities are studying night-time business operations and looking to optimize delivery schedules.

There is a growing recognition that shifting deliveries to off-peak hours can potentially reduce daytime traffic congestion, improve delivery efficiency, and decrease energy consumption and emissions [7.10]. However, other factors such as noise pollution, safe road operation and safe access to buildings, shifting work schedules and higher labor costs, and making residents comfortable with night-time delivery must also be considered. Allowing quieter EVs, cargo bikes, droids, and drones to make more efficient, autonomous deliveries after sunset, when traffic is lighter, could address noise, road safety, and labor issues. However, this could inadvertently lead to the unintended consequence of induced demand, where the convenience and speed of such night-time delivery services prompt customers to order more goods and shipments than they otherwise would!

An intriguing model that automakers, transportation companies, and mobility providers could explore is combining passenger daytime vehicle use with commercial delivery night-time operation, leveraging the unique capabilities of autonomous and electric (silent, zero-emission) vehicle technologies. A robotaxi, roboshuttle, or robopod could move passengers during the day and goods at night. This would increase the vehicle's utilization and make the unit economics of vehicle autonomy more attractive. More ambitiously, an individual's personal autonomous EV, which is used by the owner during the day, could be rented out as an autonomous delivery vehicle for a local business, logistics provider, or even a peer-to-peer courier network during the night when the owner does not require the vehicle. The vehicle could autonomously navigate to predetermined routes or destinations, autonomously recharge at centralized charging depots using automated systems when not making deliveries, and then return to the owner's residence in the morning—with minimal human intervention or operation (**Figure 7.16**).

Figure 7.16 Passenger vehicle (daytime) and commercial vehicle (nighttime) operation with a common autonomous EV.

In urban areas, fleets of small, nimble, and highly maneuverable electric, autonomous delivery vehicles (2Ws or 3Ws as well as microcars) could be deployed in concert with unmanned droids and aerial drones in dense city centers and in areas with restricted vehicle access. As a base, these microvehicles could be transported and stored in larger electric cargo vans or box trucks that operate from strategically located transportation hubs, micro-fulfillment centers, or distribution facilities in the less congested areas of the metropolitan area.

In developing parts of the world, the challenge of delivering goods is often magnified by inadequate road infrastructure and limited resources, leading to the current popularity of motorcycles for delivery service, as described in Chapter 2, Section 2.3. Fortunately, a lot can be done with software apps alone. Dynamic routing software can optimize delivery routes in real-time based on traffic conditions, road closures, and other factors while GPS tracking and accurate geocoding help identify the precise location of the delivery vehicle and customers, enabling efficient and timely route planning. Mobile apps can let customers track their orders in real time using mobile apps, as well as facilitating digital payments, reducing the need for cash handling, and improving security. Electronic proof of delivery can also be captured using GPS devices, eliminating the need for paper documentation, and blockchain can provide a transparent and tamper-proof record of deliveries, improving trust and accountability.

Drones are particularly effective when there is challenging terrain, which is often the case in developing nations and in remote areas that have poor-quality road infrastructure. Ubiquitous smartphones and data analytics can be leveraged to optimize delivery routes and improve overall efficiency, which is particularly important in areas with poor road infrastructure, where traditional logistics methods may fail. Zipline began drone delivery for medical supplies, such as vaccines and anti-venom in Muhanga, Rwanda in 2016, and by 2023 it was delivering a total of 28,754 units of blood (over four thousand gallons) to patients in Rwanda, such as women about to deliver a baby (**Figure 7.17**). The blood was delivered on average by drone within 42 minutes of when it was requested [7.11], and according to the Wharton School of Business, the blood delivery service has reduced deaths by over 50% [7.12].

Figure 7.17 Medical supplies are delivered by drone in Rwanda, provided by Zipline International.

7.6.
The Potential Role for City Government to Shape Future Mobility

Despite the potential for clear benefits of integrating public transport networks with fleets of on-demand autonomous EVs to enhance first-/last-mile connectivity, several challenges must be solved before there will be significant real-world deployment of such a comprehensive, integrated mobility model in a major metropolitan area. These include technical, regulatory, financial, and business issues.

For example, although autonomous driving capability, fleet management operations, and MaaS software platforms are developing rapidly, they are still not ready to be integrated into this future solution, however, there are good reasons to believe this can happen within the next decade. There is currently no policy framework to govern the operation of AVs as part of a public transport service, and there is institutional conservatism at public transport agencies that are more comfortable with traditional bus and rail operations. This challenge is compounded by a difficulty in sharing data and synchronizing schedules/routes with multiple mobility business providers. A lack of funding for such innovative mobility solutions has allowed well-funded robotaxi companies to dominate the discussion around future urban mobility, even if it might add to congestion and does not serve the poor and disabled as well as other solutions.

However, funding could be secured with a city policy that raises funds from robotaxi operations to support other public transport and MaaS initiatives that improve accessibility. A clear example is in issuing the licensing fee and operational permit in the first place. Cities could also limit the amount of street space available to robotaxis or they could allow robotaxis to use dedicated bus lanes, for a fee, to improve their performance. Cities can implement a variable congestion charge or VMT tax on robotaxis, based on prevailing traffic conditions, or time of day, and on occupancy level, particularly when it is empty. These charges would discourage robotaxis from adding to traffic congestion during peak hours or from driving around empty. On the flip side, robotaxi companies might sponsor city congestion charge initiatives that promise to raise traffic speeds because it would boost their productivity, and the funds raised could be reinvested into public transport.

Similarly, regulations could be imposed on where robotaxis can pick up and drop off passengers, performance standards could be mandated for response times, and zoning laws could affect the location of robotaxi parking lots and charging stations. Regulations around remote operators or maintenance staff could impact labor costs. All of these would affect operational strategy and costs, as would strict requirements for providing accessibility for the disabled, which would increase the cost of a robotaxi vehicle or create a need for two types of robotaxis. For example, the NYC Taxi and Limousine Commission uses a model where every passenger's ride includes $0.30 for a "Taxi and For-Hire Improvement Fund" that requires 50% of the traditional yellow taxi fleet to be wheelchair-accessible and imposes targets for wheelchair accessibility on TNCs (such as Uber and Lyft) in NYC [7.13]. Similarly, California's "TNC Access for All Program" collects $0.10/ride fee from each TNC ride to fund accessible transport services for people with disabilities [7.14]. Of course, if cities are too demanding in terms of taxes and fees, robotaxi operators can always choose to go elsewhere!

On the less punitive side, cities could encourage integration between robotaxis and public transport systems, potentially creating new business models or revenue streams, but this would also require mandatory data sharing, which may create additional costs for robotaxi companies even though it should lead to improved traffic management. Cities could also invest in V2I to augment vehicle sensing, which should improve safety and traffic coordination.

It is important for cities to tailor actions to their specific needs, to promote certain types of solutions that support the local quality of life and economy. For example, a compact European city may prioritize dedicated bicycle paths and robopods that work with public transport, whereas a sprawling US city like Houston, which has less public transport, may prefer to support robotaxis, and some modern, dense Chinese cities may promote V2I infrastructure support and even eVTOLs for moving people and goods [7.15].

References

7.1. Big Think, "These Maps Provide Graphic Evidence of How Parking Lots 'Eat' U.S. Cities," https://bigthink.com/strange-maps/parking-lots-eat-american-cities/.

7.2. City of Vancouver Website, "Making Streets for People Program," https://vancouver.ca/streets-transportation/making-streets-for-people-program.aspx.

7.3. Lopez, I. et al., "Mobility Infrastructures in Cities and Climate Change: An Analysis through the Superblocks in Barcelona," *Atmosphere* 11, no. 4 (2020): 410, doi:https://doi.org/10.3390/atmos11040410.

7.4. Sharma, S.N., "Basic Concepts of Transit-Oriented Development," https://track-2training.com/category/education/.

7.5. Transdev, "TransMilenio: Operated by Transdev Colombia," https://www.transdev.com/en/reseaux/transmilenio-2/.

7.6. Bladikas, A.K. and Papadimitriou, C., "Analysis of Bus Transit's Operating Labor Efficiency Using Section 15 Data," Transportation Research Record 1013, https://onlinepubs.trb.org/Onlinepubs/trr/1985/1013/1013-007.pdf.

7.7. Jeffs, J., "Autonomous Vehicle Industries Now and in 10 Years," IDTechEx, November 20, 2024, https://www.idtechex.com/zh/research-article/autonomous-vehicle-industries-now-and-in-10-years/32123.

7.8. Fortune Business Insights, "U.S. Commercial Vehicle Market Size, Share & COVID-19 Impact Analysis, By Vehicle Type (Light Commercial Vehicles, Medium Commercial Vehicles, and Heavy Commercial Vehicles), By Propulsion Type (ICE, BEV, Hybrids, and FCEVs), and Country Forecast, 2023-2030," https://www.fortunebusinessinsights.com/u-s-commercial-vehicle-market-108508.

7.9. OptimoRoute, "What Is Last Mile Delivery? Costs & How to Optimize," https://optimoroute.com/last-mile-delivery/.

7.10. Fu, J. et al., "Transport Efficiency of Off-Peak Urban Goods Deliveries: A Stockholm Pilot Study," *Case Studies on Transport Policy* 6, no. 1 (2018): 156-166.

7.11. Zipline Website, "Preventing Maternal Deaths through Faster Blood Delivery," https://www.flyzipline.com/newsroom/stories/impact/preventing-maternal-deaths-through-faster-blood-delivery.

7.12. Harriet Jeon, H., Lucarelli, C., Mazarati, J.B., Ngabo, D. et al., "Last-Mile Delivery in Health Care: Drone Delivery for Blood Products in Rwanda," January 2025.

7.13. "TNC Reform: Addressing Accessibility Concerns around Rideshare and Taxis, December 10, 2019," District 2 Atlanta, https://static1.squarespace.com/static/5cfaafb181100a00014d375e/t/5df10c869e817831826bb359/1576078470413/White+Paper+-+TNCs.pdf.

7.14. California Public Utilities Commission Website, "TNC Access for All Program," https://www.cpuc.ca.gov/tncaccess/.

7.15. Mercedes-Benz Media UK, "Pioneering Innovations for the Car of the Future: Mercedes-Benz Provides Exclusive Insights into Research Activities and Future Technologies," Mercedes-Benz Press Release, November 22, 2024, https://media.mbusa.com/releases/pioneering-innovations-for-the-car-of-the-future-mercedes-benz-provides-exclusive-insights-into-research-activities-and-future-technologies.

Chapter | 08

Rethinking Vehicle Design and Development

Vehicles, of course, are still needed even with the most progressive forms of urban design and MaaS, but what form might they take in the future? Automakers have tended to focus on horsepower and have tackled energy consumption with efficiency measures to offset "overengineering" (that is aggressively marketed), instead of supplying "effective" transport solutions to move people and goods sustainably and affordably while meeting legitimate customer needs to go wherever they want, whenever they want, as comfortably as possible. Drastically reducing energy consumption for the automobile fleet tenfold, say, will require a multifaceted strategy that encompasses not only technological innovation, but also design changes driven by a shift in how we use vehicles. This chapter looks at different approaches, considers how vehicles can learn from nature, and proposes a more effective vehicle design approach based on an ACE skateboard (or "ACE platform") that can be applied to all manner of vehicles, even boats and drones, and ranging in size from wheelchairs to heavy-duty trucks.

8.1.
Reducing Vehicle Energy Demand

The key to making vehicles more efficient is straightforward, in principle, and involves reducing the amount of energy or work needed to move in the first place and then reducing the waste of "fuel" or energy needed to perform that useful work.

The most obvious ways to reduce the energy needed to move the vehicle, particularly when accelerating, are to reduce its mass by using lightweight materials (e.g., advanced composites, high-strength steel, aluminum) and by optimizing the structure to eliminate unnecessary material, using advanced simulation and 3D printing. Exploring bio-based materials that are both lightweight and sustainable, and developing new synthetic materials with superior properties are some other examples that are being researched. Reducing the energy consumed by the vehicle and designing vehicles for easier recycling go hand in hand, as it makes little sense to use a lightweight material to reduce the vehicle's energy consumption if the lightweight material requires a lot of energy to produce in the first place!

When the vehicle is moving at a constant speed, the most effective way to reduce energy consumption is to reduce friction with the air and with the road surface, through aerodynamic drag and rolling resistance improvements. Aerodynamic improvements can involve redesigning the vehicle shape to minimize air resistance, minimizing gaps that create drag and even deploying active aerodynamic features that can adapt to driving conditions. An example of this would be an active suspension that can lower the vehicle when traveling at high speeds on smooth roads, which is made easier by the front-facing cameras that can detect the road ahead for any obstacles. Tire improvements, meanwhile, typically focus on developing ultra-low rolling resistance tires and reducing friction in wheel bearings.

Although there is some efficiency loss in the drivetrain (from the "engine" to the wheels), for most of the automobile's history efficiency improvements have tended to focus on making the existing ICE more efficient in converting the fuel's chemical energy into useful mechanical energy rather than waste heat. Recently, however, attention has shifted toward electric propulsion, either for augmenting the ICE (as in an HEV or PHEV) or for replacing it (BEV or EV, as it is commonly called), and this can be at least three times more efficient than an ICE in converting stored chemical energy (in the battery) into mechanical energy onboard the vehicle. Neither gasoline nor electricity is a primary energy source since they do not occur naturally, but converting and refining petroleum into gasoline is more than twice as efficient as burning coal or natural gas to create electricity for charging a battery. The goal, of course, is to replace dependence on one fuel source (petroleum) with many possible energy pathways to make electricity. This can include renewables, which can also be cleaner for the environment, both globally and locally. Automotive research is also considering hydrogen fuel cells as the source of electricity for the electric motor because this could extend vehicle range, but it faces stiff competition from battery improvements that promise higher energy density and can enable lighter vehicles, as mentioned in Chapter 4, Section 4.3. Although highly efficient by ICE standards, research continues into making electric motors and power electronics even more efficient as this will also reduce battery size (for a given range) and heat losses in the wiring harness. Wheel motors may be one way to achieve this as was suggested in Chapter 4, Section 4.4.

After reducing vehicle weight and increasing powertrain efficiency, systems that consume the most power, such as HVAC, will yield the most opportunity for energy savings. Other electrical systems, such as lighting and infotainment, consume relatively little power. Heating and cooling passengers more efficiently can be an effective way to increase vehicle range (or to downsize the battery for a given range). Auxiliary methods of HVAC innovation that do not require forced air are needed, especially because quieter HVAC operation also improves comfort and supports quiet EV operation. Some examples include ambient floor heating, heated steering wheels, and heated or cooled seats and headrests. Advanced insulation could reduce heating and cooling needs, phase-change materials could be used for thermal energy storage, and solar panels can lower peak cabin temperatures when the vehicle is parked, reducing ancillary HVAC loads. Toyota claims its bZ4X solar panel roof adds >1000 miles annual range [8.1], which could be exploited to reduce battery size and vehicle weight. Incorporating high-efficiency solar cells into vehicle exteriors and developing transparent solar cells that can cover the glazing surfaces are active areas of development and are covered in more detail in Section 8.3.

Solar capture is a form of energy harvesting, which refers to the process of capturing and converting ambient energy sources to supplement energy from the vehicle's battery, thus extending the range of the EV and lowering operating costs. The most well-known form of energy harvesting is regenerative braking, which is particularly effective for vehicles that frequently decelerate and stop, such as city cars. Other types of energy harvesting include generating electricity onboard the vehicle from kinetic energy dissipated in the shock absorbers (e.g., piezoelectric conversion) and waste heat from the propulsion system (thermoelectric conversion), but neither has yet been commercialized.

Piezoelectric materials have the potential to convert some of the vibration energy into electricity when shock absorbers compress and expand during a car's motion over a bumpy road surface. Unfortunately, most road shocks generate small forces compared to what piezoelectrics are typically designed for, and this produces a low-energy output in the milliwatt range. Research into increasing the conversion efficiency is focused on matching the road bump frequency to the piezoelectric material's specific vibration frequency and optimizing the placement of the piezoelectric material in the shock absorber system. Improving the robustness and durability of piezoelectric materials in the harsh vehicle environment is another active area of research. It is possible that the application might shift toward supporting low-power car features like sensors or interior lights, rather than trying to contribute significantly to the car's main power source. A heavy, off-road military vehicle (where cost may be more easily accommodated) will be a better candidate for energy harvesting than a small, lightweight vehicle that operates mostly on relatively smooth city streets. This is because more energy can be harvested from a larger battery's waste heat and from the more aggressive shock absorption needed on bumpy surfaces. Moreover, the incremental cost associated with adding energy capture is likely to be a higher fraction of the total vehicle cost for the smaller, lighter EV because most hardware costs per watt fall as power increases.

However, as the technology matures and economies of scale are achieved, costs should fall and the benefits could become more pronounced.

Thermoelectric materials have been looked at for energy recovery from ICEVs, which generate a lot of waste heat. Roughly one-third of the gasoline's energy content performs useful work moving the vehicle, a similar amount is converted into heat that is removed by the radiator, and another third is heat that goes out the exhaust system. Because the exhaust is the highest temperature on the vehicle, this is where thermoelectric devices are most likely to work the best, because their performance depends on maintaining as large a temperature difference as possible. However, even with high exhaust temperatures (around 700°C), they generate a relatively small amount of electricity, and increasing the temperature difference between the exhaust system and the ambient may require additional heat exchangers that must withstand harsh underbody environments and could undermine the catalytic converter's ability to clean exhaust emissions. For waste heat recovery from EVs (batteries and electric motors), the challenge is even greater because these components are nowhere near as hot as engine exhaust. One area of research into improving thermoelectric conversion efficiencies is focusing on promising advanced materials that can increase electrical conductivity while lowering thermal conductivity. However, even if the technical issues can be solved, the amount of electricity generated could still be modest and might be difficult to justify for most commercial applications, given the added cost and complexity.

Beyond these measures of weight reduction, powertrain efficiency and energy harvesting, there are several other opportunities for reducing energy consumption based on actual, real-time usage rather than on the actual vehicle design. With improved navigation systems that can predict the route ahead, and with electrified powertrains that enable regenerative braking, there is potential to consume less energy by limiting hard acceleration and braking events, thus smoothing out the driving profile. In the longer term, vehicles may be able to coordinate their actions via V2X communications, discussed in Chapter 4, Section 4.8, reducing the amount of time spent idling at traffic lights and platooning or coasting with other vehicles traveling in the same direction. Implementing driver feedback systems to encourage or gamify energy-efficient driving is a way to incentivize drivers, particularly as it also tends to improve safety and may lower insurance premiums. Admittedly, this will require a shift in consumer expectations and behavior, as well as supportive policy environments, and for both reasons may be implemented first in China.

8.2.
Nature's Lessons for Future Vehicle Design

Nature has evolved highly energy-efficient systems, and by studying them, engineers can gain valuable insight into designing more energy-efficient vehicles. In the realm of structures, for example, nature often uses strong but lightweight materials and structures to provide support without adding unnecessary mass. Examples include hollow bird bones, bamboo, and the Venus flower basket, which uses delicate glass-like materials to form a strong but flexible structure. Michelin's Tweel, a nonpneumatic tire, is inspired by the honeycomb design often found in nature, such as in spiderwebs and bone structures, that can bend without breaking. Human bones can self-repair and regenerate after minor fractures, which is guiding researchers to study how to create automotive coatings and surfaces by activating embedded polymerization catalysts that can self-heal from microcracks and chips. In short, various automakers are studying nature's designs, using computational tools, to optimize vehicle structures to be extremely strong yet lightweight. This could lead not only to innovative vehicle structures but also to associated manufacturing optimization, because nature encourages circularity, sustainability, and regenerative design.

Many birds and fishes have evolved streamlined body shapes that minimize drag as they move through the air or water. The unique body shape of a boxfish allows it to glide through water very efficiently with minimal drag, and Mercedes-Benz has studied it to design more aerodynamic vehicle shapes that reduce wind resistance and improve fuel efficiency. Other examples include a shark's skin that has tiny scales for reducing turbulence and is inspiring researchers to develop similar vehicle coatings. Owls, meanwhile, can fly silently because of specialized feather adaptations that manipulate air turbulence, and Toyota is looking at how this design could reduce wind noise and turbulence around side-view mirrors. The Shinkansen bullet train's nose design was inspired by the kingfisher; its long, tapered beak allows it to dive from the air into water with minimal splashing, and similarly, the Shinkansen train can enter tunnels at high speed without creating an unpleasant sonic boom, leading to more efficient and faster trains. The principles of aerodynamics observed in birds and flocks, such as the V-formation used by wild geese to create an ascending air current, are likely to be adopted in the future design of coordinated automobiles to reduce drag.

Nature has evolved efficient energy storage systems in organisms like electric eels and certain bacteria, while some deep-sea creatures produce light efficiently (**Figure 8.1**). By studying these systems, scientists may be able to develop better batteries for EVs and more efficient, bioluminescent-inspired lighting systems for vehicles.

Figure 8.1 Two examples of nature-inspired automotive design. a) Kingfisher, b) Shinkansen "bullet" train, c) spider web and d) non-pneumatic car tire.

When it comes to energy conversion and storage, it is well known that plants can respond to changes in their environment, such as light or water availability, and that animals can change their gait for different terrains and speeds. In a similar way, vehicles equipped with sensors and adaptive systems can optimize their energy consumption based on real-time conditions. A vehicle's suspension system must absorb and release significant amounts of energy. Some animals, like kangaroos, store energy in their tendons when landing, which is then used to power their next jump, and a cheetah's spine acts as a shock absorber during high-speed runs. The way their spine flexes and stores energy could inspire more efficient suspension systems that not only cushion the ride but store energy from bumps and dips in the road and release it to aid acceleration or to power onboard systems. In addition, trees are remarkably efficient at harvesting energy from the wind through their flexible structure, and the way leaves and branches move could inspire designs for multiple small energy harvesters in a vehicle's suspension system, each contributing to the overall energy collection. Similarly, insects like the cockroach have multiple legs that all contribute to energy-efficient movement, and this distributed approach could inspire suspension designs where multiple smaller energy harvesters work together throughout the vehicle.

Animals, such as elephants, use their large ears to regulate body temperature, while sweating and panting are other natural cooling mechanisms. Can vehicles incorporate more effective passive cooling systems inspired by such natural adaptations? Or consider the lotus leaf's microscopic surface pattern that makes them extremely hydrophobic, causing water to bead up and easily run off, helping to remove dirt. Nissan has mimicked this "lotus effect" to create self-cleaning automotive glass, mirrors, and body panels that actively resist water, fog, and dirt accumulation. By applying these principles to automobile design, it may be possible to reduce the need for windshield wipers and associated energy load and to passively improve visibility during adverse weather conditions.

Learning from nature does not stop with the "hardware" because the brain and "software" are increasingly becoming sources of inspiration for future automotive solutions, especially as vehicles become software-defined and the limitations of conventional software approaches become clear. Nature and biology can also serve as a rich source of inspiration for software design and development.

AVs, in particular, can draw inspiration from nature to emulate the strategies used by animals, based on acute sensory perception, to detect potential collisions. By using cameras, radar, and lidar, AVs can identify objects in their path, similar to how bats rely on echolocation to navigate their environment. Moreover, animals exhibit adaptive decision-making to avoid collisions, and by incorporating machine learning and reinforcement learning algorithms, AVs can learn to make real-time decisions based on environmental cues, rewards, and penalties, just like how animals adapt their behavior to avoid collisions. In terms of actuation, AVs can be developed to perform evasive maneuvers akin to an animal's instinctive response. For example, evasive-steering assist and autonomous navigation systems can help AVs to avoid imminent collisions by steering away from obstacles, mirroring the natural agility of animals in avoiding potential threats. And finally, by using reinforcement learning, AV software can be trained to avoid collisions by being penalized for collisions and rewarded for successfully avoiding them, fostering a safety-oriented approach to software development. This is reminiscent of how animals learn to navigate their habitats safely.

In the realm of software development for AVs, drawing inspiration from the brain's computational processes can significantly improve machine learning, as discussed in Chapter 4, Section 4.6. For example, emulating the densely interconnected neurons of the brain, software development can leverage neural networks to mimic human learning by adjusting the strength of simulated neural connections based on experience, supporting more efficient and human-like AI and allowing AVs to process complex data and make intuitive decisions, similar to the brain's neural processing capabilities. Brain-inspired computing technologies, such as neuromorphic computing, offer a different way to process information and with extreme energy efficiency. They leverage the brain's approach of co-locating memory and processing, encoding information differently, and employing massive parallelism, and this paves the way for more efficient, less power-hungry compute systems.

Finally, recent developments have shown the potential of using human brain activity data to guide machine learning algorithms, infusing the training process with insights from working brains to create more consistent and biologically inspired learning models. In summary, these approaches hold the potential to advance the capabilities of AVs and pave the way for more human-like and efficient AI in auto-mobility.

8.3. Solar Power

Nothing seems more natural than solar power, and the case for its integration into vehicle design is building each year as performance improves and as costs come down. Coupled with right-sized vehicles, the case for solar panels on the vehicle itself is becoming compelling. Solar-powered EVs can charge during the day, improving their energy efficiency, reducing the dependence on fossil fuels, and reducing strain on the electric grid.

Some production vehicles already have a solar panel that is integrated onto the roof. The first-generation Nissan Leaf had a small solar panel to trickle-charge the 12-V battery that powered some accessories while the current Toyota Prius Prime plug-in hybrid offers a solar roof option that contributes to the car's battery and helps power ventilation, shown in **Figure 8.2**.

Recently, Hyundai Motor Company unveiled a hybrid electric Sonata sedan that has a solar roof-mounted system that can charge around half of the battery's capacity per day, which can increase the vehicle's travel distance by up to 700 miles annually if the vehicle is charged six hours a day [8.2]. Not to be outdone, Toyota has worked with NEDO to integrate Sharp's highly efficient solar cells on a plug-in hybrid Prius' roof as a vehicle concept. These solar cells are just 0.03 mm thick and generate 22.5% solar battery cell conversion efficiency on the Prius PHV [8.3].

In some cases, not only the roof but also the doors, hood, pickup box tonneau cover, tailgate, and trunk can be used for mounting solar panels to increase solar energy collection, albeit with lower efficiency than the roof, and some solar panels are now flexible enough that they

Figure 8.2 Toyota Prius PHV with solar panel roof, 2016 Paris Auto Show.

can be molded directly into composite or glass to become a structural element. For example, a Chinese company, Teijin Limited, has recently developed a curved solar roof with their Panlite polycarbonate resin glazing that meets aerodynamic, aesthetic, and strength targets for automotive applications [8.4].

Several startups, such as Aptera Motors, Lightyear, and Sono Motors, have begun to develop ground-up solar-powered EVs with sufficient solar panel coverage to significantly extend the vehicle's range. Aptera Motors Corp. plans to produce a three-wheeled solar vehicle with a carbon-fiber composite body covered in three square meters of solar cells, adding as much as 45 miles of solar range each day, which is more than the average number of miles driven per day in an automobile.

Lightyear, a Dutch company, is developing the Lightyear One, which has five square meters (approximately 55 square feet) of solar cell coverage protected by strong, double-curved safety glass (**Figure 8.3**). A conductive back sheet allows all the electrical connections of the solar cells to be put on the back of the solar panel, which increases the solar collection area by 3%, and also improves the aesthetic appearance. The solar surfaces are claimed to capture enough sunlight to provide more than half of the vehicle's annual mileage needs, reducing operating costs and trips to the charging station and putting less strain on the electric grid during the summer heat. Because Lightyear One has been designed to be highly efficient, roughly 50% more than a typical comparable-size EV, and because the solar panels are roughly 20% more efficient than most commonly used solar panels, sunlight can add approximately seven miles of range with each hour of solar charging [8.5].

Figure 8.3 Solar panels are best placed on horizontal surfaces and can be mounted on the hood and trunk to increase surface area.

Mercedes-Benz is researching five micron-thick photovoltaic films that could even be applied over the vehicle's entire body, with an area of approximately 10 square meters. This could generate enough energy to cover most of the average daily driving needs even in temperate places like Stuttgart, and supply all the average daily driving needs for cities in sunnier climates [8.6].

In short, solar panels are no longer merely a means to power a fan blower to pull down cabin temperatures when the vehicle is parked in the sun. They are now being seriously considered for extending the EV range and for burnishing the environmental image of the automaker and vehicle owner. In the future, the surface area available for solar collection might increase if transparent photovoltaics become available and, perhaps, used in sunroof or rear window applications; but, for now, transparent solar cell technology has a significantly lower power output than typical opaque solar cells. With L4/L5 AVs, the NHTSA requirement for at least 70% visibility through the windshield should no longer be relevant and this would open up the large windshield area for solar collection and enable the type of cheaper, "opaque" solar panels commonly used today. Currently, windshields have a rake angle of 20 to 35° above horizontal and there may also be more of an opportunity with AVs to alter this and increase solar energy collection.

Some engineering challenges still remain, however. Maximizing solar collection in suboptimal weather and shading conditions for a double-curved roof surface will be essential to get the most out of the solar panels. Because solar cells are in series, the shaded part can reduce the total output from the whole panel, so predicting the orientation of the sun's rays on the vehicle roof, based on location and time of day, will help with planning where to park the vehicle to avoid shading and to maximize solar collection. This could be easier with autonomous fleet operation where the vehicle may be routed to specific parking areas as part of an optimized system. Ford has demonstrated a C-Max Solar Energy Concept that tracks the sun's movement and uses a Fresnel lens (magnifying glass) to concentrate the sun's rays onto the solar panel roof [8.7].

For passenger vehicles, another challenge is finding ways to use the solar energy once the battery is fully charged. One approach is to use the solar electricity to run fans to circulate air inside the cabin to avoid 150°F peak cabin temperatures. For commercial vehicle fleets, there may be additional opportunities because solar panels could, for example, reduce auxiliary power needs for a refrigerated truck or keep the battery from discharging while running power tools for a contractor.

The amount of solar radiation in a place like Arizona or India is approximately 6000 kWh/m^2/year, meaning a two square meter panel, typical of a car roof, can generate 2.4 kWh each day if it is 20% efficient (**Figure 8.4**). For a typical car that may average four miles/kWh (not a truck!), this could provide as much as ten miles daily range on solar power alone, saving perhaps $1/day. This size solar panel may cost around $250 while current vehicle installation costs can add up to $2000, depending on the car model and labor costs. Finally, the cost of power electronics, such as a charge controller and inverter, for charging the battery from the solar panel, can bring the total cost of integrating solar panels into a vehicle roof up to $3000. This would mean it could take 3000 days or over eight years to pay back the upfront cost. This explains why solar car roofs are not widely seen today.

Figure 8.4 Twenty percent efficient solar panels mounted on a typical two-square-meter vehicle roof can generate more than two kWh/day in many parts of the world.

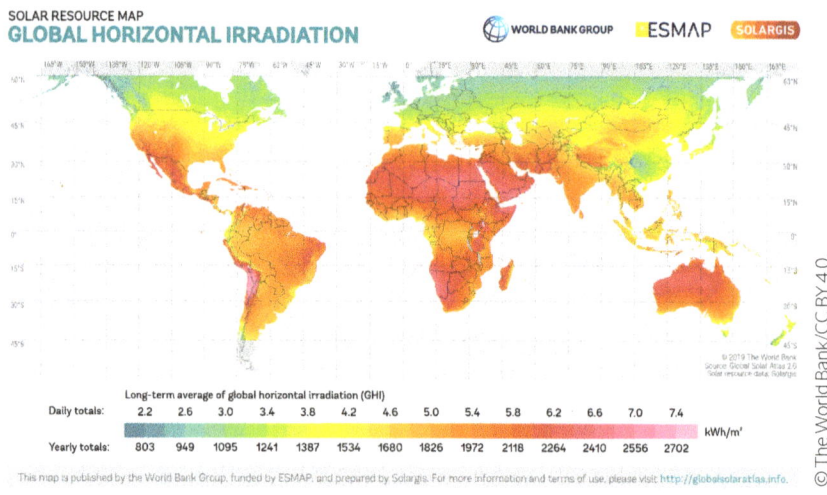

At present, a payback period on energy savings like this is not attractive for most buyers, but as solar panel manufacturing continues to improve and production scales up, the cost of solar panels is expected to decrease, and as more automakers integrate solar panels into their vehicle designs, the installation process will benefit from mass production and become much more cost-effective. Moreover, advancements in power electronics technology, driven by EV competition, should also decrease the cost of power electronics for battery charging.

In addition to reducing cost and improving the flexibility of solar panels, future research must increase the efficiency of solar energy conversion. This will likely require advancements in both the solar panel and automotive power electronics, and may also require innovations in solar tracking (by tilting the solar panel and/or positioning the vehicle) to maximize solar collection across the day and minimize shading. It is reasonable to expect improved efficiencies, reaching 25%, by 2035 (**Figure 8.5**).

Figure 8.5 Declining $/W trend for solar photovoltaic modules, with prices extrapolated to halve again by 2030 to around $0.10/W [8.8].

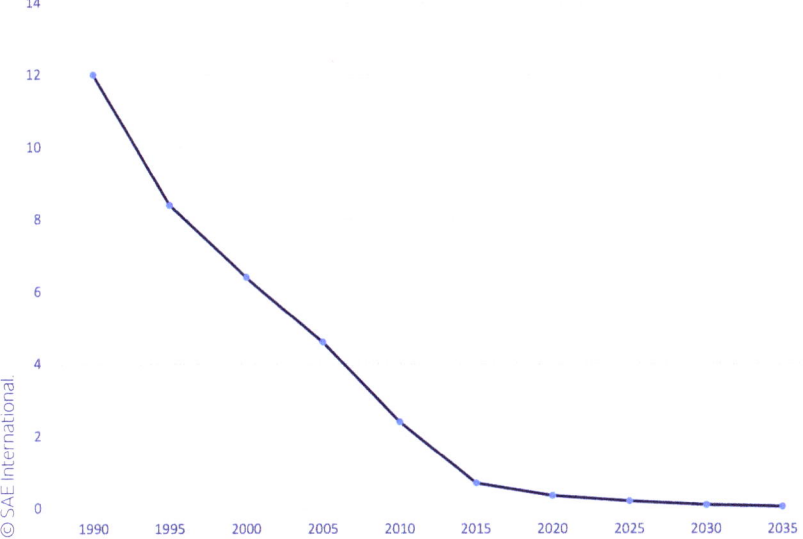

By 2035, solar panel module prices may be a third of current prices and have an expected 25% efficiency. This would mean that the same-size solar panel as used in the previous automobile example could provide 12 miles of range per day, or nearly 4400 miles/year, and with an installed vehicle cost of approximately $1000. In such a scenario, the price of solar-roof-generated electricity will be $0.12/mile, comparable to many ICEVs, but only during the first two years of operation. Afterwards, it would be free forever!

8.3.1.
Right-Sizing the Vehicle

Of course, if the EV was as efficient as a "golf cart" or the type of microvehicle described in Chapter 3, Section 3.2, then the daily solar range with a two-square-meter roof panel could reach 40 miles, which exceeds the average distance driven each day. In this scenario, the solar panel could provide all the energy needed to move the vehicle and for $0.07/mile in the first year, with free electricity for the rest of the vehicle's life, making it very compelling, particularly for a fleet owner. This brings us to the topic of vehicle right-sizing, and as discussed in Chapter 7, Section 7.2, encouraging the use of smaller, more efficient vehicles for moving people and goods in daily use that are appropriate for urban environments, in particular.

As shown in Chapter 2, Section 2.1.2 even if it is difficult to quantify exactly how much current vehicles are overengineered for "90%" of usage, most people will acknowledge that cars offer horsepower and acceleration far beyond what is needed for everyday driving. Although ADAS features can be beneficial for safety, they are not strictly necessary because nobody "needed" them until they were recently introduced. Towing capacity and off-road features, common in SUVs and trucks, may never be used in daily urban driving, while features such as premium sound systems and panoramic sunroofs are considered luxuries for many drivers.

Right-sizing the vehicle can have a tremendous "ten-fold" impact on cost, weight, and efficiency, as was shown in Chapter 3, **Table 3.1**. An electric tuk-tuk, increasingly common on Indian city streets, can cost and weigh ten times less than an electric pickup truck, and the battery alone can be ten times smaller and lighter, with huge implications for reducing supply chain, national security, environmental and human rights concerns. For fleets, with well-defined duty cycles and usage, there may be no situation where the "excess" functionality of a pickup truck or car is needed, and if it is needed occasionally, then it might be less costly to have just a few such vehicles in the fleet rather than the entire fleet, despite the extra complexity this can cause. Microvehicles, which tend to be micro-EVs are quickly gaining in popularity, particularly in dense Asian megacities where pollution, congestion, and parking are major transport issues. Because these vehicles tend to be driven a short distance, they only need small batteries and can be cheaper than ICEVs, especially when running "fuel" costs are included.

Electric two- and three-wheelers (E2W/E3W) typically retail for less than $4000 in Asian markets (whereas in developed markets they can cost up to $10,000). Vehicle manufacturers in developing countries have adapted their products to cater to local market needs and preferences. For instance, companies like Bajaj Auto and TVS Motor Company in India have historically developed low-cost, fuel-efficient motorcycles and scooters tailored for urban commuting and rural transportation, and the transition to electric mobility is happening rapidly in many places in the developing world. Companies like Ather Energy and Ola Electric in India and Gesits and Viar in Southeast Asia are leading the way in developing E2W/E3W that are well-suited for urban environments and have less environmental impact. The rise of shared mobility services in developing countries is also promoting the use of right-sized vehicles for personal mobility and last-mile delivery. These services often leverage two- and three-wheelers, optimizing resource utilization and reducing congestion.

Many applications and use cases across different sectors can be effectively and economically served by "right-sized" EVs with driving ranges that are short compared with ICEVs. Fleet operators, businesses with predictable local travel patterns, and operations confined or "geo-fenced" to a specific geographic area are particularly well-suited to adopt EVs optimized for a lower total cost of ownership over the vehicle's lifetime, rather than focusing solely on upfront costs, performance metrics, or brand prestige. Beyond just urban commuting and last-mile delivery, the developed world could explore the use of E2W/E3W for other purposes like emergency services, postal/courier services, and even light commercial transportation. Additionally, the development of some dedicated infrastructure like motorcycle lanes, parking facilities, and battery charging or swapping stations for E2W/E3W is likely to be crucial to support their wider adoption. This could involve reallocating road space, implementing traffic management systems, and incentivizing the use of these smaller vehicles through policies and subsidies.

One size up from E2W/E3W are microcars (commonly known as neighborhood EVs or NEVs in the US and quadricycles in Europe). As with expensive e-bikes, these vehicles may currently cost as much as entry-level cars even though they lack the safety content and have reduced performance and range. The high cost is likely because the market for this type of vehicle is almost nonexistent in developing Asian economies and relatively small in developed countries, where they may be seen as "unsafe" and limited to low-speed retirement communities in the US, for example. The Renault Duo and Citroen Ami are signs that European automakers are beginning to see a role for this type of vehicle (**Figure 8.6**). Nissan has also decided to begin distributing a similar vehicle, the Spanish-built Silence Nano, in the UK.

Figure 8.6 The Citroen Ami, a quadricycle, can be operated without a driver's license.

Compact "Kei" car models, originally developed by Japanese automakers like Daihatsu, Honda, and Suzuki to comply with domestic tax and parking incentives for cars under a certain size, could find growing demand internationally (**Figure 8.7**). Unlike quadricycles or NEVs, these compact EVs are designed to meet basic FMVSS and are capable of highway operation because they can travel at speeds up to 70 mph, but they could have a modest range of up to 100 miles on a single charge, which can help make them more affordable. They can be optimized for specific fleet applications like last-mile delivery services operated by logistics companies or retailers, geofenced vehicles for self-contained communities like gated neighborhoods or university campuses, and local goods transportation within urban areas for small businesses or vendors. However, investments in public charging infrastructure, workplace charging facilities, and residential charging solutions may be necessary to support the widespread adoption of these specialized, short-range EVs.

Figure 8.7 Although Japan's Kei cars are as tall as a pickup truck they are well suited to dense urban centers because they are approximately three-fourths the width and length of a typical US crossover.

As discussed in Chapter 7, Section 7.2, a compact, limited-range EV that meets car safety standards could be the optimal solution for a robopod fleet that connects riders to suburban public transport stations in developed cities, while in combination with a viable subscription service, it could also become the household's solely owned vehicle.

8.3.2.
Right-Sizing the Battery

In households with higher incomes and the financial means to own several vehicles, EVs could initially gain traction as a "backup" car that ends up doing most of the driving! The current generation of EVs, while offering more than 200 miles range on a single charge from advanced lithium-ion battery packs, may not yet have the capability to fully replace a household's primary vehicle that would be used for long-distance travel, road trips, or hauling/towing needs. However, as the household's second vehicle, an EV can serve as an excellent and economical urban runabout for the vast majority of trips (**Figure 8.8**).

Figure 8.8 The effect of doubling EV range on cost of ownership and the environment.

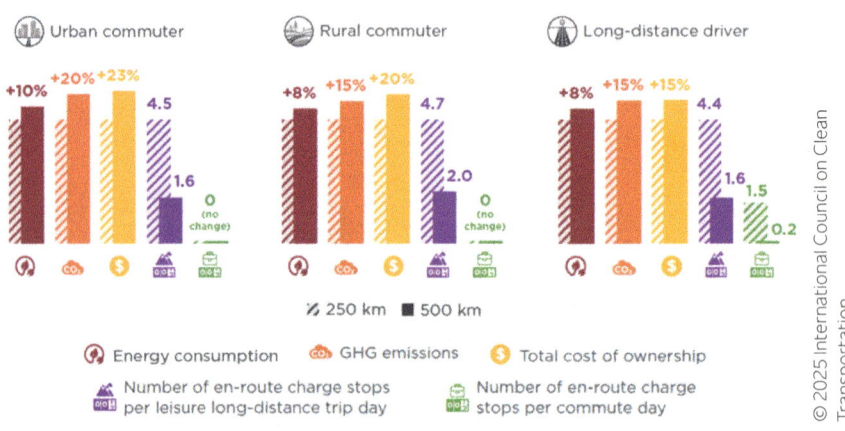

The International Council on Clean Transportation (ICCT) has analyzed how battery size affects EV energy consumption [8.9]. They simulated a compact EV with four different battery sizes over a one-year period, allowing for a driving range under World Harmonized Light Vehicles Test Procedure (WLTP)-type approval conditions of between 275 and 784 km (roughly 170 to 490 miles). Their simulation considered three types of users: a typical urban commuter, a rural commuter, and a driver taking frequent long trips with a higher annual mileage.

The study concluded that when the real-world range doubles from 250 to 500 km, the higher vehicle weight causes energy consumption to increase by nearly 10% and life-cycle greenhouse gas emissions increase by almost 20% because a larger battery also takes more energy to make. The higher vehicle purchase cost and the higher electricity operating costs cause the cost per kilometer or mile to go up 15% for the long-distance driver and 23% for the urban commuter. In summary, a larger battery is only needed for long-distance trips, which occur on fewer than seven days a year on average. For all user types, a smaller battery combined with more frequent fast charging saves money because the high cost of fast charging is more than offset by the lower vehicle purchase price and lower energy consumption.

Fleets with predictable routes and schedules can determine if switching to EVs makes financial sense [8.10] and can calculate the optimal battery size for their vehicles, but for individual consumers an EV purchase is likely to be a more difficult decision. Consumers buying EVs with "right-sized" batteries may be making a logical decision, but from an emotional perspective, they will also need reassurance that the EV charging infrastructure is "fit for purpose." For now, most EV owners tend to be affluent and have another ICEV or HEV with more than 300 miles of range, meaning that a 200-mile range EV with home charging can be used for local driving, while for occasional, longer trips, the household's ICEV/HEV can be used. However, as EV adoption takes off, it may become the household's only vehicle and so there must be a solution to address easy EV charging. This means it needs to be readily available and working at all times, offer a variety of charging speeds and payment methods, and provide a pleasant user experience. In Chapter 4, Section 4.2, it was noted that some EVs, such as Ford's Lightning Pro Power, have a 220 V generator outlet that can power a home, and California has recently introduced a bill requiring most EVs that will be sold in 2027 to have V2G bidirectional charging capability. With a common communications protocol, this would allow vehicles to directly charge each other, and because there will be many more EVs than public charging stations, this should increase the ubiquity of charging, and alleviate range anxiety, particularly for apartment dwellers who might not have access to charging. Recharging another EV generates revenue for the owner when the vehicle is parked, but it may also reduce battery life as the battery will be cycled more.

If charging needs can truly be met, as is being done in China with the ultrafast DC charging breakthroughs announced by CATL and BYD, for example (Chapter 3, Section 3.3) then it should be possible to market EVs to consumers even with 200 miles of range, making them more affordable to purchase and operate because they would be lighter and cost less to recharge. Solving the charging challenge to enable smaller batteries will significantly increase national security and reduce environmental concerns and automaker vulnerability to battery materials scarcity. The potential for reducing international conflict and creating more business certainty should encourage national governments and the automotive industry to prioritize charging infrastructure investments, but mechanisms will need to be created for assigning a monetary value or government incentive to this approach.

Another approach to reducing range anxiety for consumers was discussed in Chapter 7, Section 7.4, and requires a novel mobility subscription service. Several premium OEMs currently offer subscription packages that allow customers access to many vehicles but not on the daily or weekly basis that would make this a practical solution for single-EV owners.

For fleets that operate with a well-defined route and schedule, there are many strategies that can help to reduce battery size, and with a well-defined operating area, there is even the possibility to optimize battery chemistry as well. For example, for electric buses that operate on fixed routes and can charge at the same location at the end of the route, wireless EV charging (WEVC) could allow buses to have a smaller battery while still meeting daily driving needs for performing the route multiple times. The business case for this depends on ensuring a sufficient number of buses can exploit the same WEVC charging pad at different times of the day to maximize its utilization. In this way, the aggregate fleet battery cost savings and higher efficiency (because of a smaller, lighter battery) can justify the WEVC infrastructure and vehicle pad cost.

Or consider EVage, an Indian EV startup, which is developing and manufacturing commercial electric trucks and buses in microfactories. It selected Toshiba's LTO battery over LFP and NMC chemistries because it offered the lowest cost solution despite having a higher cost/kWh [8.11]. This was possible because the battery needed to store only 15 kWh and was sufficient for a 50 km range. Fast charging twice a day and slow charging overnight could allow the vehicle to meet daily driving needs but it requires that the battery can be cycled 10,000 times while operating safely in a hot climate all year round. This combination of requirements (fast charging, long cycle life, safe performance at high ambient temperatures, and well-defined charging protocols) led to a small LTO battery providing the lowest overall cost solution with the highest vehicle efficiency and payload capability.

Tires are optimized for a variety of parameters, but there can still be a justification for winter tires that have a special formulation to assist traction on icy roads. A similar approach can be applied to batteries as well. A fleet operator might swap batteries out and maintain two sets of batteries, although this adds operational complexity and overhead. Conversely, in locations where the temperature is reasonably constant throughout the year, the battery chemistry can be optimized for that location because fleets, unlike personal vehicles, may have a well-defined geographic operation. Different vehicles in the fleet may also have different batteries if they operate in different climates. In other words, because of their tighter operating domain and organized maintenance operations, fleet EVs may have an opportunity to optimize battery selection rather than be constrained to using the same batteries as personal vehicles. Moreover, the well-defined daily usage patterns and charging infrastructure availability could enable lower-energy-density, less expensive battery chemistries, such as Na-ion batteries, to be viable, especially for E2W/E3W and micro-EVs.

As discussed in Chapter 4, Section 4.3, a competing electrification solution is the hydrogen FCEV, particularly for the Class 8 heavy-duty truck segment. For applications with minimal downtime, high payloads and range requirement FCEVs may be a good solution and could reduce supply chain risk with battery sourcing. However, the emerging fleet deployment of battery electric trucks with charging infrastructure installation could make having additional FCEV servicing and infrastructure complexity unappealing to existing battery electric fleets in the future.

Putting a value on powertrain-related innovation like this has, historically, been justified by automakers needing to meet CAFE regulations to avoid fines. For example, automakers developed a "$/lb saved" metric for ICEVs, and because efficiency improvements can reduce battery mass and cost, a similar cost calculation can be made for EVs. The difference for EVs is that a risk premium should also be added to this calculation because reducing the battery size and the amount of battery materials needed will lower the automaker's vulnerability to battery material availability and price spikes. If automakers quantitatively assign a benefit for this risk reduction to their business, then more efficiency innovations could be justified.

The smart approach to reducing future battery materials supply chain risk must be to "right-size" the batteries and reduce the demand for the battery materials. This is true on environmental, human rights, and national security grounds. Procurement and supply organizations can treat this risk like an insurance agency and incentivize decision-makers inside their companies to find creative ways to tackle battery right-sizing. This could be achieved by putting a $/lb saved premium on risk reduction, pursuing solar panel integration, applying different batteries for different use cases and locations, viewing charging infrastructure and mobility subscriptions as competitive differentiators, and so on.

8.4. Car-Free and/or Low-Speed Zones

The future of vehicle design will be driven not only by technology "push" and customer "pull" but also by public policy.

As will be discussed further in Chapter 9, Section 9.5, when cities implement policies that either restrict cars from entering certain zones or enforce speed controls for all vehicles inside the zone, then innovation can skyrocket. As shown with Barcelona's Superblocks in Chapter 7, Section 7.1, vehicles will still be needed for moving people and goods (for emergencies, last-mile deliveries, assisting the disabled, during bad weather, etc.). However, if the zone is enlarged it can become practical for dedicated vehicles to operate that may not need to meet FMVSS requirements and will have far more modest range and speed requirements, consistent with geo-fencing in pedestrian-rich environments. This could enable a much wider choice of structural materials, such as recycled plastics, which promote a circular economy (Chapter 9, Sections 9.2 and 9.3), as well as a much smaller battery that might obtain most or all of its daily energy needs from a solar panel roof, depending on the climate. It would also open up more battery chemistry options and help battery materials supply chains become more resilient. The combination of low vehicle mass, speed, and range means that a 5 kWh battery (~$500!) could achieve up to 100 miles driving range, and with fast charging once during the day, it should be possible to meet daily driving needs for operating in the city center. For moving people and goods around inside the city center, this is a far more affordable and environmentally friendly solution than using conventional automobiles.

It may be politically challenging to ban cars from the city center, but there may be less drastic ways to stimulate the development of micro-EVs that can be made from locally available materials or even be made from recycled plastics, thereby reducing landfill waste. For example, in 2020, BMW introduced an eDrive Zone feature on PHEV models, which allowed the car to automatically switch to EV mode in city centers, by relying on GPS to identify designated areas for geo-fencing. In a similar GPS-enabled manner, intelligent speed assist could be designed to automatically control vehicle speeds to the designed speed limits in city centers which will improve road safety for pedestrians and cyclists. This would also make it easier to introduce new types of vehicles, restricted to city center operation, that do not have to add the cost and mass needed for meeting car crash safety standards. According to the IIHS, most drivers accept their vehicle providing an audible and visual warning when they exceed the posted speed limit [8.12], but it is clearly another matter if the vehicle controlled or governed the speed! Acceptance of speed control is likely to be higher in city centers, where traffic speeds tend to be slow and where there are more pedestrians and cyclists, than on highways, especially if these vulnerable road users are also residents in the city center.

With typical speed limits of just 20–30 mph and range requirements potentially as low as 50–100 miles per charge in these restricted urban zones, EVs intended for such duty cycles can utilize lighter-weight construction materials and more cost-effective components that are optimized for lower mass and overall manufacturing cost, as opposed to the heavy steel body structures designed primarily for occupant protection in high-speed collisions.

Rather than relying solely on traditional vehicle segments and form factors like sedans, SUVs, pickup trucks, or vans, new mobility platforms and architectures can be purposefully optimized and tailored for different transportation applications beyond just personal mobility. For example, micro-EVs could be designed like GM EN-V (Chapter 3, **Figure 3.18**). This vehicle has a skateboard enabling a low center of gravity but with a seating position that is at eye level with pedestrians. It has front entry access to and from the street curb, instead of exiting the vehicle onto road traffic. It also has an exterior design that prioritizes safety for vulnerable road users like pedestrians, cyclists, and children, which is better suited for dense urban environments, areas around schools, community centers, or retirement facilities. These small vehicles can integrate well with public transport, both physically and electronically, because compactness makes it easier to park them at stations and their autonomy, connectivity, and electrification make it easier to redistribute vehicles for pickup, drop-off, and automatic or wireless recharging.

Ultimately, the success of more affordable micro-EVs in a car-free or speed-controlled urban environment will depend on factors such as appropriate safety regulations that are tailored for micro-EVs, public acceptance and adoption of these lightweight micro-EVs, as well as their integration with the urban infrastructure and public transport systems. With the appropriate regulatory framework, consumer demand, and infrastructure support, the absence of stringent crash standards could pave the way for affordable and sustainable micro-EV solutions in future car-free city centers. Although politically difficult to restrict movement of any kind, this approach would improve road safety and could even be politically popular as it can create high-quality local employment in the design, development, and manufacture of new city vehicles, which will be discussed more in Chapter 9, Section 9.4.

It is important to note that this vision of localized, low-speed micro-EVs, whether autonomous or not, is likely to coexist with personal car ownership and choice. Rather, it enables established automotive companies as well as new mobility startups, in partnership with cities, to provide innovative mobility solutions that can effectively augment public transport options in areas where conventional automobiles are not practical from a technical, economic, or urban planning perspective, essentially filling mobility gaps that personal vehicles struggle to address. Beyond public transport, other use cases may include short-range transportation within gated communities, pedestrian zones or residential areas, personal ride services, and small-scale delivery and logistics needs in areas with vehicle access restrictions.

8.5.
The ACE Platform "Reference Design" Applied to Future Automobiles

The barriers to entry for making cars are falling because EVs are simpler to develop, with less need for "exhaustive" emissions testing (pun intended) and ICE calibration. Meanwhile, the importance of vehicle production for creating high-value jobs and exports was well understood by China, and other developing countries, such as India, Saudi Arabia, and Vietnam, are hoping to copy this playbook. As discussed in Chapter 6, this has coincided with traditional automaking countries, such as the US, Japan, and Germany, being slower to transition away from ICEVs than China.

The automobile is undergoing its biggest change since it was invented in the 1880s. Future cars will drive themselves and communicate with each other to avoid collisions and reduce congestion, and they will be electric to address energy security and environmental concerns. These three trends are often called autonomy, connectivity, and electrification (or "ACE"). Right now, during this awkward transition period (2015–2025), car accidents are up after declining for 40 years. Many pin the blame on our "need" to text on the phone while driving. But the same smartphone technology and architecture may also hold the key to morphing the car into a "smartphone on wheels" (**Figure 8.9**).

Figure 8.9 The "smartphone on wheels" analogy has caught on, despite some notable differences between vehicles and smartphones, such as safety and reliability.

The automobile as a "smartphone on wheels" is an imperfect analogy that has been made many times in recent years because both are electric-powered, are connected, and include a wide variety of sensors (cameras, microphones, accelerometers, etc.). The "smartphone on wheels" moves the vehicle away from a traditional mechanical device and into a consumer electronics product that is increasingly software-rich, with a growing consensus on the need for a single compute platform, common in laptops and phones. This paradigm shift is upending how the auto industry designs, makes, and sells cars in the future, which should excite consumers and scare the auto companies, many of whom may be powerless to prevent their own demise.

To see how this affects the auto industry, let us briefly look at how cars are made today. Simplistically, a car is made from many independent mechanical systems (frame, brakes, steering, glazing, air-conditioning, etc.). Since the 1970s, many vehicle systems have become more and more electrically operated and software-controlled. The suppliers for each system bundle software with the physical hardware, and the automaker integrates and develops software that links all these supplier parts together. After the software has been integrated, the vehicle is sold to the consumer as a final product. It has a fixed performance, hopefully, until the vehicle's end of life. Because the hardware is initially expensive, it is often first applied to luxury vehicles and then trickles down to mainstream brands and models a few years later.

Now let us look at how future vehicles will be designed. With batteries underpinning future vehicles, we have what is called an electric "skateboard." This is the natural platform replacing the "horse and carriage" that subliminally drove engine placement in front of the passengers around 1900. The skateboard, shown in Chapter 3, **Figure 3.3**, includes all the running gear (frame, motors, batteries, brakes, steering, suspension, wheels, and tires), and it creates much more design freedom for the "coach" or passenger compartment. A flat floor means seats can be moved around. Wheelchair access is easier, as is loading and unloading parcels. The length and width of the skateboard can be tailored for different markets, especially if wheel motors are used (as covered in Chapter 4, Section 4.4). Dense city centers with limited space need shorter vehicles, for example. With the right autonomy and connectivity hardware, the skateboard can be used for autonomous pods, robotaxis, shuttles, and autonomous goods delivery vans. This fusion of autonomous, connected, and electrified hardware inside a skateboard could be considered an ACE platform, which is effectively a large smartphone on wheels. The sensors, such as cameras, radars, and lidar, may be mounted onto the coach body, but the underlying software that ties them together to create autonomous operation will be part of the ACE platform.

Since it can underpin many types of vehicles, the ACE platform can use one set of common, high-performing hardware. As pioneered by Tesla, this creates economies of scale across multiple vehicle lines instead of the common practice of developing and making individual vehicles having their own bespoke hardware. The ACE platform saves on the cost of parts and nearly all the development work and manufacturing equipment can be reused for other vehicles spun off the same platform. With a common set of hardware across multiple vehicles, the key to providing different performance and experiences will be through software and a steady stream of OTA update improvements. The ACE platform developer may create these apps or curate those developed by third parties, but app developers will want to work on the most powerful ACE platform because this drives the most customer pull for software enhancements.

Companies will still be able to differentiate their ACE platforms not just with software but also by using more capable or powerful hardware that can turbocharge future vehicle performance and the customer ride experience. Some examples include wheel motors and active suspensions (Chapter 4, Section 4.4), cabin monitoring (Chapter 4, Section 4.7), and V2X (Chapter 4, Section 4.8). These innovations all provide multiple benefits to the end customer as well as to the manufacturer and can also be adjusted and tuned with software.

An ACE platform comprising a skateboard with additional hardware to support autonomous connected EVs and some level of "future-proofing" (such as wheel motors, active suspension, cabin monitoring, and V2X) is well-suited to support a variety of applications for moving people and goods and can easily be adapted for both types of applications (**Figure 8.10**). The same vehicle could even be converted, during night hours, from moving people to moving goods, as mentioned in Chapter 7, Section 7.5. It can be tailored to navigate the tightest alleys and streets, and to occupy as little space as possible and reduce the real estate needed for storage overnight. It will provide the smoothest ride for fragile goods and the best ride experience for passengers. It will ensure the autonomous operation is safe and smooth, like a train and not a bus.

No single company, with the possible exception of Tesla, currently has all the competencies needed to develop the ACE platform, and so a partnership between a technology company (likely with smartphone expertise) and an auto company (either an automaker or a major automotive supplier) will be required. Even with such a partnership, it is expected that each partner will need to be software and hardware integrators and work with various suppliers. Whoever develops this ACE platform first will be in a pole position to supply it to anyone else, particularly if early sales can jumpstart data collection to feed into future versions of the ACE platform. The list of customers could include other automakers, mobility operators, commercial ride-hailing and goods delivery fleets, public transit agencies, car dealers and rental agencies, communities, and any aspiring business entrepreneur wanting to penetrate the mobility space.

Figure 8.10 ACE platform and a possible partition of responsibilities.

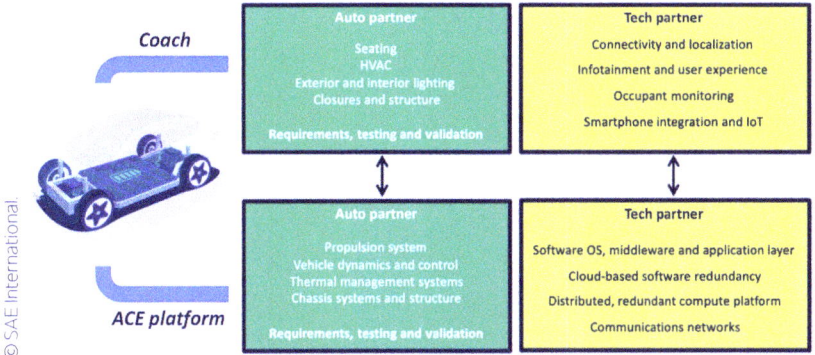

This is where the analogy with the smartphone goes even further because the ACE platform is effectively a reference design, and the smartphone industry, unlike the auto industry, is very familiar with this concept. In a similar manner, an automaker (e.g., Tesla), could potentially license an EV design. This could be the current design if it is licensed to a noncompetitor or, perhaps, the most recent design if it is licensed to a competitor.

During the late 2000s, China was concerned it would fall behind other countries, such as South Korea, in developing smartphones and wanted to create more domestic competition for the established, foreign players. Qualcomm created reference designs for the smartphone's "running gear" and licensed them to startup Chinese smartphone companies. These companies then competed with each other on the customer-facing side (styling, user interface, price, marketing, etc.). This program was so successful that within a few years there were many Chinese smartphone makers competing in China. Not surprisingly, as the Chinese smartphone industry has grown in recent years, more and more of the hardware is now being developed inside China by Chinese companies.

Companies like Foxconn, which manufactures Apple iPhones, and Magna Steyr, which manufactures some limited-volume vehicles for several major automakers, are also well-placed to execute this vision, and Foxconn has already debuted several complete vehicle reference design concept vehicles that can be modified by potential business customers [8.13]. In December 2024, CATL showcased its "Bedrock Chassis" skateboard, which it intends to supply to other companies. This has the battery cells directly integrated into it so that 85% of the vehicle's collision energy can be absorbed, compared with about 60% for a traditional chassis. The system is also designed to disconnect the high-voltage circuit within 0.01 sec and complete the vehicle's residual high-voltage energy discharge within 0.2 sec, which is claimed to be a new industry record [8.14]. These innovations make it the state-of-the-art for EV safety, according to CATL. Automakers, such as Tesla, could potentially license their EV designs as well. This could be a current design if it is licensed to a noncompetitor or, possibly, a previous generation version to a competitor.

This type of reference design model was first applied to smartphones. During the late 2000s, China was concerned it would fall behind other countries, such as South Korea, in developing smartphones and wanted to create more domestic competition for the established, foreign, players. Qualcomm created reference designs for the smartphone's "running gear" and licensed them to startup Chinese smartphone companies. These companies then competed with each other on the customer-facing side (styling, user interface, price, marketing, etc.). This program was so successful that within a few years there were many Chinese smartphone makers competing in China. Not surprisingly, as the Chinese smartphone industry has grown in recent years, more and more of the hardware is now being developed inside China by Chinese companies.

Because there are many companies in the automotive and mobility space who do not have the economies of scale and R&D prowess to create an ACE platform, there is an opportunity for an ACE platform to be developed and licensed, at scale, to many new and existing players around the world. As with the smartphone reference design, many companies could compete by focusing on customer-facing aspects (design of the coach, passenger experience or goods movement, price, mobility service, etc.). In one sense, this gives customers much more choice and increases competition among vehicle builders and mobility service providers because it dramatically lowers the barriers to entry for new players. On the other hand, it reduces competition among ACE platform developers and may create a winner-takes-all scenario that has been seen across many new businesses, including the smartphone. Should this happen, we will see the demise of the traditional automotive industry.

8.6.
The E-Kit as an ACE Platform for Nonmotorized Vehicles (NMVs)

For many other applications and parts of the world, the most common type of vehicle is an NMV. Encompassing anything with wheels that can be pushed, pulled, or pedaled, this class of low-speed vehicles includes tricycles, handcarts, and wheelchairs to name just a few examples that may be relevant in urban delivery, agriculture, and healthcare use cases.

There is significant potential for synergies and shared components between the AV and EV solutions described so far, and these low-speed use cases spanning sectors such as healthcare, agriculture, and logistics. Healthcare facilities and hospitality venues (hotels, airports, convention centers), warehouses and manufacturing plants, as well as large office or residential complexes, share many core mobility needs that depend on efficiently and cost-effectively moving people and goods within the site area without noise or pollution, fostering electric operation. Where labor costs are high, autonomy will be pursued as well.

The idea of a skateboard ACE platform is conceptually easy to visualize for light-duty vehicles (passenger cars and trucks) as well as for heavy-duty vehicles (buses and commercial trucks). It can also be applied to 2W/3W (**Figure 8.11**) and NMVs as a platform, albeit not taking the shape of a physical skateboard. For these vehicle applications, the ACE platform could be replaced by a standardized, modular set that comprises electric propulsion (motors, inverters, and battery packs) and control systems (electric brakes and steering) housed in a box, perhaps with wheels, which would serve as the foundation for developing a wide range of both indoor and outdoor power-assisted electric mobility solutions across disparate market segments. A cheap autonomy system comprising sensors (e.g., cameras), connectivity, processing, and software could be developed and be functional in simple or controlled environments, perhaps even using the customer's smartphone and docking it to the same vehicle. In a sense, this "e-kit" can be thought of as a mini-ACE platform. This standardization of key components into an e-kit ACE platform could offer scale economies in production, reducing per-unit costs (**Figure 8.12**). Of course, the wheels and tires will need to be adapted to the different use cases (larger ones are needed for agricultural than for indoor applications, for example), but the core components of the e-kit might be largely the same.

Figure 8.11 Projected EV sales (light-duty vehicles, heavy-duty vehicles, 2W/3W) [8.15].

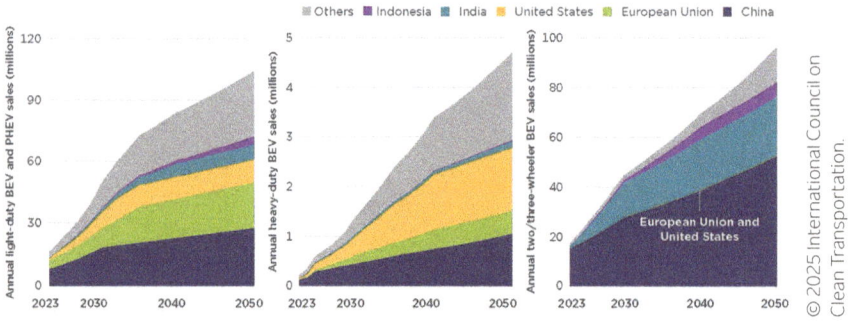

Figure 8.12 A universal e-kit (or a mini-ACE platform) that can be retrofitted to a wide variety of NMVs to support applications, such as healthcare, agriculture, and urban delivery.

In healthcare facility applications, for example, this e-kit could provide electric power assist for wheelchairs, hospital beds, gurneys, and meal and linen carts to reduce workplace injuries and the number of staff required (**Figure 8.13**). It could also be used to lift heavy patients off the bed. A more advanced version of the e-kit could move autonomously and be shared among multiple "robots" within the hospital environment (e.g., carts for moving patients, pharmaceuticals, medical equipment, meals, and laundry) and autonomously move itself to recharge wirelessly when necessary. A shared e-kit could make financial sense versus outfitting each bed, gurney, and wheelchair with separate e-kits because they do not need to move around most of the time. This is particularly true in less affluent environments, such as rural hospitals, that may not be able to afford to have every bed motorized.

Figure 8.13 Healthcare applications for an e-kit to provide power assist for moving hospital beds, wheelchairs, and lifting a hoist.

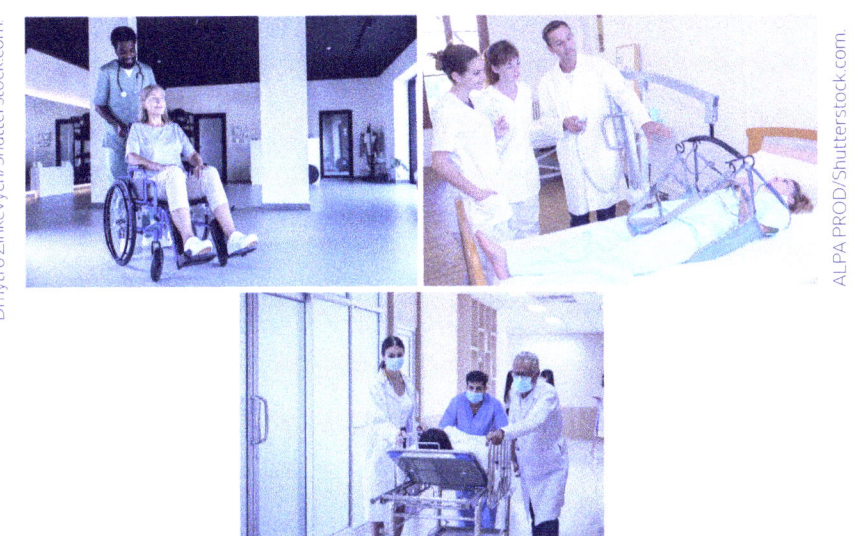

The hospitality sector represents another fertile area for the e-kit with applications such as a guest luggage delivery robot, mobile concierge service kiosk, and autonomous food and laundry service rovers for room service delivery. Other potential e-kit applications include autonomous mobile robots for order fulfillment in warehouses, materials transport inside a manufacturing facility using automated guided vehicles (AGVs), autonomous delivery rovers operating in large office parks or gated residential communities, and electric baggage tugs or pushback tractors operating within airports.

All these applications and many more could leverage the same core e-kit components. Beyond the scalability benefits, a key advantage of developing and producing a standardized e-kit is that it streamlines procurement, interoperability, and maintenance of these electrified mobility solutions across diverse environments and duty cycles. Shared training, tools, and maintenance facilities can be leveraged. Furthermore, an innovative business model enabling shared usage and collective investment into pools of these standardized e-kits across multiple businesses, facilities, campuses, or even municipalities could be explored. This would increase asset utilization rates, lowering capital expenditure requirements for individual organizations while ensuring optimal deployment based on actual demand using intelligent software to manage and coordinate the shared e-kit fleets.

The e-kit can also be applied anywhere in the developing world, where the cost of a new vehicle is often beyond reach, but where retrofitting to a familiar and existing NMV could be an attractive solution. This approach could help to make electric-powered mobility more affordable upfront while also ensuring low operating costs. In many parts of the developing world where there is no access to the electric grid, the only viable source of energy is solar and so the e-kit would also need to include solar panels to charge the batteries that will provide power assist and portable power (**Figure 8.14**).

Figure 8.14 M-Kopa, in Kenya, has pioneered the use of an innovative business and financing model encompassing solar panels to power LED lights and TVs in rural, off-grid locations. a) Powering LED lights, b) powering a TV.

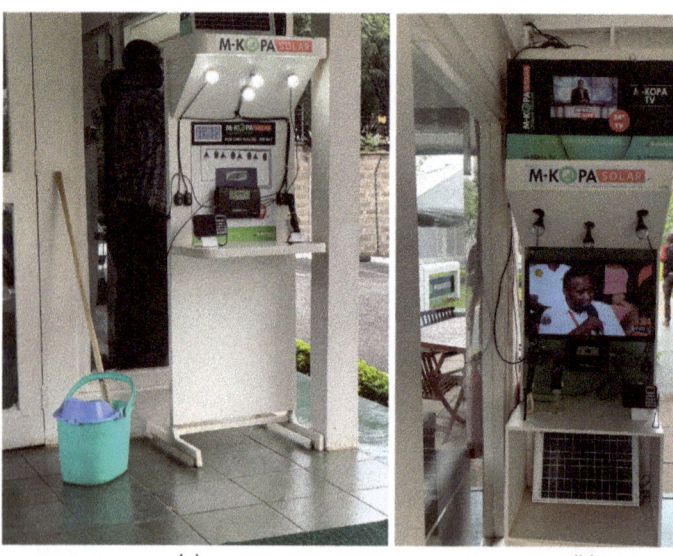

Retrofitting existing NMVs, such as tricycles or carts, with solar panels should be less costly than purchasing an all-new solar EV because much of the vehicle has already been paid for and the user is familiar with it. Here again, the potential for a shared e-kit exists if one considers attaching it to a wheelbarrow or cart operating inside a farm for moving mulch around and then reattaching it to a tricycle for transporting harvested produce to a nearby market. Many parts of the developing world receive abundant sunlight, and by harnessing solar energy, users can eliminate their dependence on fossil fuels, which is particularly beneficial in areas where fuel prices are volatile or where access to fuel is limited. Solar panels are easy to maintain, which is advantageous for users in developing countries who may not have access to extensive repair services. A solar retrofit could be tailored to various types of NMVs, including rickshaws and carts used for transporting goods, and this flexibility allows for a wide range of applications, from personal transportation to commercial use. The solar e-kit could even be retrofitted to river boats (common in Southeast Asia), reducing water pollution caused by outboard motors as another benefit, and with drones that are particularly useful over rainforests in tropical regions.

While retrofitting is generally cheaper than buying new EVs, the upfront costs of solar panels and installation can still be a barrier for some users in developing regions. However, as solar technology continues to decrease in price, this barrier is gradually diminishing. At scale, the price of an e-kit should be under $500/kWh, given that solar panels and lithium-ion batteries are the two most expensive components and should each cost around $100/kWh. This assumes $1/W solar panel cost ($1000/kW) with ten hours of charging each day, and the potential for reusing lithium-ion battery modules from used EVs (or continued cost reduction in new lithium-ion batteries). If a 2 kWh e-kit is sold for $1000 and has a minimum of a three-year life (1096 days), then the daily cost of ownership should be roughly $1. Given that many people in sub-Saharan Africa currently pay at least $0.10 to charge their cell phone and that a smartphone battery has around 10 Wh of energy, it may be possible to use the e-kit to charge ten phones, cover its daily lease cost, and still leave 1.9 kWh of energy for providing power-assisted mobility to make it easier to transport people and goods to the market, workplace, hospital, or school.

Volunteering Experiences in Mali

During my volunteering experiences in Nana Kenieba, a rural Mali village of around 700 people, with Professor Mark Bryden (of Iowa State University, ISU) and some of his students, I was struck by a couple of observations. The first was when I learned that an entrepreneurial villager was using the solar panel charging station that had been developed at ISU to charge lead acid started batteries and was renting these batteries to people in the village. One person used the battery for LED lighting, so he could work at night to make clothing garments for sale. LED lights are safer, cleaner, and cheaper than using kerosene, and the extra work allowed him to increase his income. Another used the power to grind corn for villagers, a manual task that often involves women pounding the corn heavily with a wooden pole. Other valuable applications include charging cell phones and pumping water from underground.

The village of Nana Kenieba, shown in the map (top left) on page 287, is located about 50 miles southwest of Bamako, the capital city of Mali.

Sustainable and Affordable Mobility for All **285**

The roof-mounted solar panels (top image) were used to provide LED lighting for night work (bottom image).

Roof-mounted solar panels for charging lead acid batteries were also used to grind corn.

The second came a few days later when I walked approximately ten miles between three villages to help repair some solar-powered water pumps. Underground water is more predictable than the rains and much cleaner than the open well water but it requires electric pumps to draw the water up. I combined these ideas and got to thinking whether an entrepreneur could rent out a battery-powered vehicle and not just the battery. Over the course of several years, working with several companies and universities in the metro Detroit area, a few different "Afreecar" vehicle prototypes have been built and tested.

Sustainable and Affordable Mobility for All

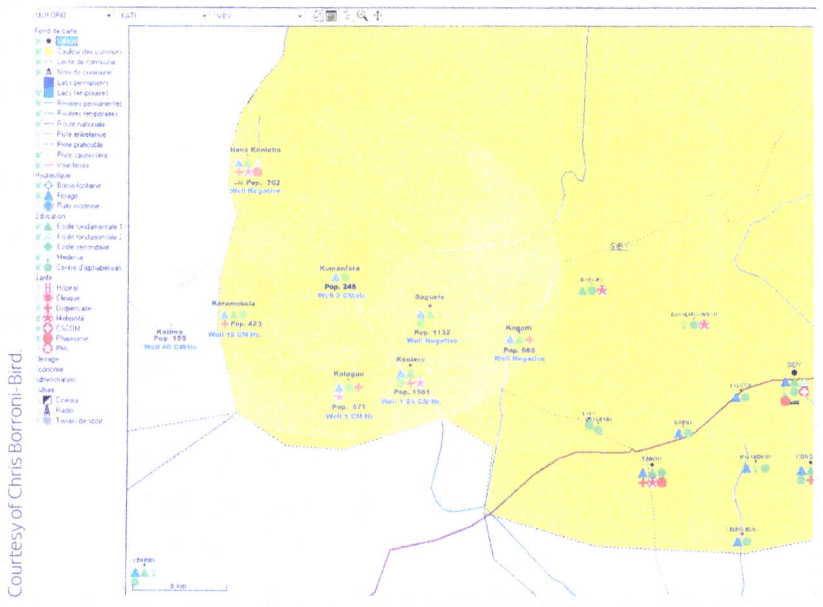

A typical distance between villages is around five miles. The main method of transport is walking, plus a minibus ("sotrama") that comes once a week to allow people to sell produce and crafts at the market.

Solar-powered vehicle prototypes that have been built include an e-trailer that could attach to any bicycle and an e-cart.

Sustainable and Affordable Mobility for All **289**

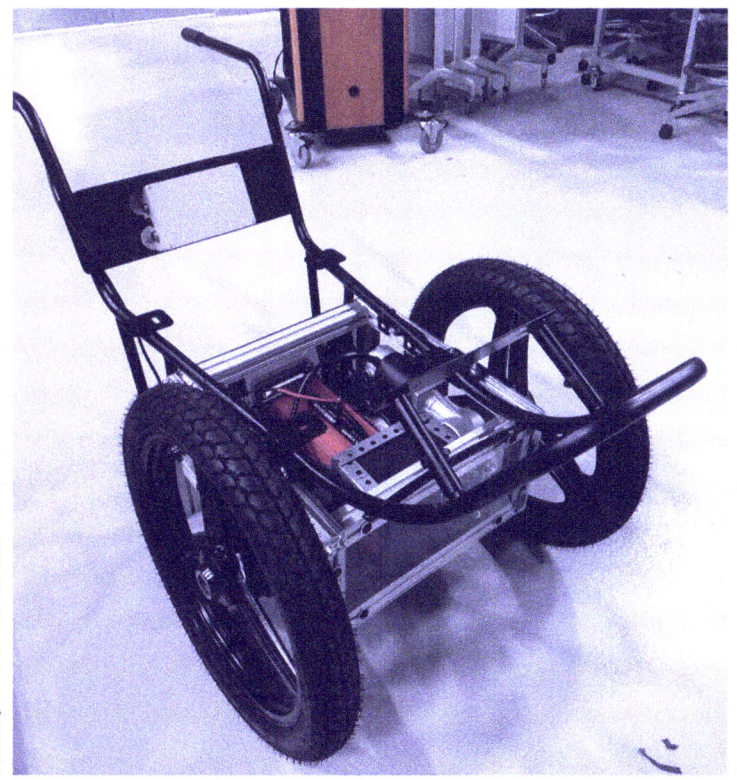

The e-kit is designed to attach to a wide variety of NMVs, such as a hand cart, and provide both power-assisted transport and mobile power for charging electrical devices, like cell phones, water pumps, ventilators, etc.

The Afreecar idea has evolved from trying to determine the best vehicle design and requirements to engineering an e-kit that could be attached to the NMVs that are already being used, such as hand carts, wheelbarrows, and tricycles. The current goal is to demonstrate the utility of the e-kit in a pilot program and to define the business case for how it may be profitable for the manufacturer while being priced at a point that represents good value for the user. Once both of these work packages are complete, the vision is for entrepreneurs and governments around the world to support local production and usage of the e-kit to create valuable employment and support economic development.

References

8.1. Hawkins, A.J., "Toyota's New Electric SUV Has a Solar Roof and a Steering Yoke Like Tesla," The Verge, October 29, 2021, https://www.theverge.com/2021/10/29/22752539/toyota-bz4x-electric-suv-steering-yoke-solar-roof.

8.2. Halvorson, B., "2020 Hyundai Sonata Hybrid: What to Expect from Its MPG-Boosting Solar Roof," Green Car Reports, April 27, 2020, https://www.greencarreports.com/news/1127957_2020-hyundai-sonata-hybrid-what-to-expect-from-its-mpg-boosting-solar-roof.

8.3. Fuscaldo, D., "Toyota to Test New and Improved Solar Power Cells Vehicle System," Interesting Engineering, July 5, 2019, https://interestingengineering.com/transportation/toyota-to-test-new-and-improved-solar-power-cells-vehicle-system.

8.4. Plastics Engineering, "Charging Ahead: Teijin Develops Solar Roof for Electric Vehicles," May 27, 2023, https://www.plasticsengineering.org/2023/05/charging-ahead-teijin-develops-solar-roof-for-electric-vehicles-000034/.

8.5. Yekikian, N., "To Solar Panels and Beyond: Lightyear One Is an EV that Loves the Sun," MotorTrend, April 20, 2020, https://www.motortrend.com/news/lightyear-one-solar-powered-car-details-price-pictures/.

8.6. Mercedes-Benz Media UK, "Pioneering Innovations for the Car of the Future: Mercedes-Benz Provides Exclusive Insights into Research Activities and Future Technologies," Mercedes-Benz Press Release, November 22, 2024, https://media.mbusa.com/releases/pioneering-innovations-for-the-car-of-the-future-mercedes-benz-provides-exclusive-insights-into-research-activities-and-future-technologies.

8.7. Szondy, D., "Ford's C-MAX Solar Energi Concept Sports Rooftop Solar Panels," New Atlas, January 3, 2014, https://newatlas.com/ford-cmax-energi-concept/30279/.

8.8. Dahlmeier, U., "Empirical Approach Shows PV Is Getting Cheaper than All the Forecasters Expect," *PV Magazine*, December 5, 2023, https://www.pv-magazine.com/2023/12/05/empirical-approach-shows-pv-is-getting-cheaper-than-all-the-forecasters-expect/.

8.9. Poupinha, C. and Dornoff, J., "The Bigger the Better? How Battery Size Affects Real-World Energy Consumption, Cost of Ownership, and Life-Cycle Emissions of Electric Vehicles," ICCT, April 9, 2024, https://theicct.org/publication/bev-battery-size-energy-consumption-cost-ownership-lca-ev-apr24/.

8.10. Ford Media Center, "New Ford Pro Tool Runs the Numbers to Show Where Electric Makes Cents," https://media.ford.com/content/fordmedia/fna/us/en/news/2024/10/17/new-ford-pro-tool-runs-the-numbers-to-show-where-electric-makes-.html.

8.11. Energetica-India Website, September 27, 2024, https://www.energetica-india.net/powerful-thoughts/online/inderveer-singh.

8.12. IIHS HLDI Website, "Most Drivers Would Be OK with Anti-speeding Tech in Vehicles, Survey Shows," June 12, 2024, https://www.iihs.org/news/detail/most-drivers-would-be-ok-with-anti-speeding-tech-in-vehicles-survey-shows.

8.13. Hon Hai Technology Group (Foxconn), "HHTD 2024 Unveils Two Reference Design EVs, Extending Range of Electro-Mobility to Midi Bus, LMUV," Foxconn Press Release, October 8, 2024, https://www.foxconn.com/en-us/press-center/press-releases/latest-news/1433.

8.14. "CATL Launches the Bedrock Chassis that Withstands 120 km/h Impact without Catching Fire or Exploding," CATL Press Release, December 24, 2024, https://www.catl.com/en/news/6343.html.

8.15. Li, E., Bieker, G., and Sen, A., "Electrifying Road Transport with Less Mining: A Global and Regional Battery Material Outlook," ICCT Report, December 15, 2024, https://theicct.org/publication/ev-battery-materials-demand-supply-dec24/.

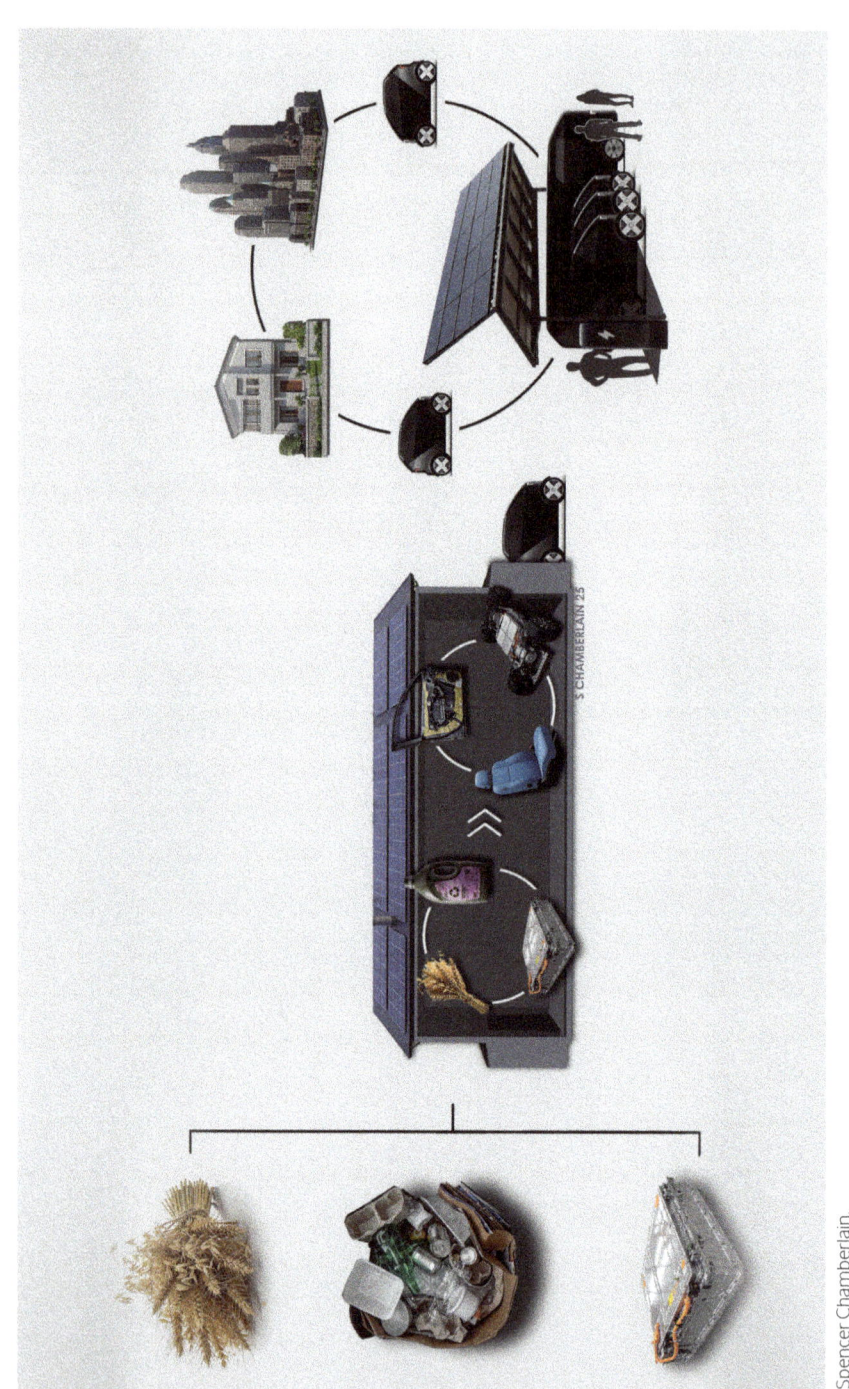

Chapter 09

Rethinking Vehicle Materials and Manufacturing

9.1. A Paradigm Shift

Nearly all passenger cars and trucks sold today are designed and developed with conventional automotive structural materials and mass production in mind. There are several reasons for this, but the key motivator is to generate economies of scale to drive down costs to encourage more sales and higher profits.

It was not always this way. Before World War I, automobiles were considered a luxury item, and each one was unique, handcrafted, and built by skilled craftsmen. This led to slow production rates and high costs so that a typical car could cost a hundred times a typical weekly wage. Henry Ford realized that standardization could dramatically reduce costs and that vehicle assembly required a sequence of steps because it is logical to install the engine before the body panels restrict access for the workers, and to paint the body at the end of the process.

These insights led to the assembly line concept, where the car moves between specialized stations with workers focused on specific tasks.

Large stamping presses are needed for the body panels, paint shops are required, and welding the body together requires precision accuracy and tooling. Each of these processes scales because the initial cost for the equipment is spread over a larger volume, and the marginal cost to make each additional part or car is low. For example, a paint shop's air handling and curing ovens will cost the same whether 100 cars need to be painted or 100,000! Although automated welding robots and stamping presses have high upfront costs, their operating costs are low.

Having all the processes in one factory, or to be more exact, a factory complex, lowered the transportation costs and helped to ensure consistent quality control across all steps, while continuously optimizing the production flow and managing inventory (**Figure 9.1**). The downside to mass production, of course, is that it requires companies to make a big bet on future vehicle sales to justify the costly investment in manufacturing a new vehicle model. Manufacturing processes and technologies need to be continually improved to stay competitive, because a new vehicle model might be made from different materials and require a new type of powertrain to meet customer or regulatory demands for higher fuel economy, and these changes will also affect the manufacturing process.

Figure 9.1 A typical automotive manufacturing plant complex (Ford, Valencia).

Since Henry Ford's time, mass production has become increasingly capital-intensive because massive stamping presses for steel body panels cost millions of dollars, robot welding lines require precise positioning and calibration, and paint shops need to be controlled environments that have increasingly stringent emissions controls. Assembly lines need a balanced production flow, standardized worker training, and specialized quality inspection controls equipment. Because the automakers do not make all the vehicle components themselves, their purchasing and manufacturing operations must manage complex supply chains from all around the world. Suppliers must also invest in their own dedicated tooling, expect volume commitments from the automaker to justify lower pricing, and supply their systems just in time to the automaker, with minimal inventory management.

At the same time, vehicle development costs have also mushroomed for a variety of reasons, including a need for increasingly sophisticated crash-testing (e.g., front, side, rollover) and precise testing and manufacturing of safety systems (e.g., airbags, crumple zones). Vehicles also require extensive simulation and testing for powertrain efficiency and emissions, NVH (noise, vibration, harshness), aerodynamics, and extreme climate conditions that the vehicle could be exposed to once it is purchased. (The kind of geo-fenced EVs described in Chapter 8, Section 8.4 will not require as much testing because they produce no emissions and operate at low speeds in a well-defined climate on well-defined surface streets.)

Mass production has allowed consumers to buy vehicles that are both affordable and reliable, and fierce competition has driven automakers to become more efficient, investing in automation and sometimes putting them at odds with their unionized workforce. For most of the twentieth century, the automotive industry created a "virtuous" circle, whereby large capital investments helped to lower per-unit costs and increased the overall size of the market, which justified further capital investment and competition. At the same time, it can also be argued that "putting people on wheels" stimulated economic development and raised standards of living to allow more consumers to trade up every few years with a new vehicle purchase.

Consumers expect consistent quality and rely on dealerships for any maintenance and repair work, which requires brand building and scale. Because finance and insurance organizations, automotive supply chains and workers, regulatory agencies, and the supporting infrastructure for repairs, maintenance, and refueling are all tuned in to the current automotive business paradigm and have optimized their processes for it, any alternative model will likely face resistance because it may lack an established financing model or supply chain, and it may require different worker skills, regulatory changes, and infrastructure upgrades. And, of course, it may face a skeptical end consumer.

However, driven by various trends, such as urbanization, goods delivery, and subscription models, there are strong reasons to believe that the market for new types of low-speed, right-sized vehicles, described in Chapter 8, Section 8.4, will grow significantly and that a different method to produce them may make more sense than the traditional automotive mass production approach. This is because low-speed urban EVs, for example, have a much simpler vehicle architecture with less need for a complex structure that is designed for high-speed crash protection. It also uses a simple electric propulsion system with no exhaust emissions, and there is less waste heat and noise to manage.

Lower production volumes are a problem for traditional automobile manufacturing but might not be as much of an issue for vehicles that are much simpler to build and might be sold to fleets, who are more concerned with the total lifecycle cost of ownership than with the initial purchase price. The vehicles could even be sold to municipalities and form part of a public transport system or a public–private MaaS partnership. As part of this tighter integration between vehicle production and city or regional governments, it is also possible that vehicle production might be localized to create employment, as described in Section 9.4.

In this paradigm, different manufacturing approaches may be possible and could include using composite materials that do not require huge presses, modular assembly instead of a continuous assembly line, and the elimination of paint requirements by using powder coating or unpainted materials. With lower production volumes, simpler automation can be used at lower speeds. However, some traditional aspects of mass production will remain, such as a need for consistent quality control, integrated supply chains, and proven assembly techniques. Finding the sweet spot between the efficiency of mass production and the flexibility of localized production could lead to a hybrid model where some components or modules, perhaps even the ACE platform, come from automakers, contract manufacturers or auto suppliers, and benefit from automotive economies of scale, while some modules and final vehicle assembly are performed locally.

The rest of this chapter deals with how this new paradigm can support a circular economy by encouraging the use of recycled and natural materials in place of traditional automotive materials, how this new manufacturing paradigm can be implemented with supportive government policies, and how the automakers can complement their existing business model and play a constructive role in this new paradigm.

9.2.
The Circular Economy

Approximately 80 million automobiles are produced around the world each year. Assuming the average automobile weighs around 1500 kg, then this means that over 100 million tons of automobile material is produced each year. This far exceeds the aggregate amount of any other consumer electronics product or appliance and highlights just how important the auto industry is as a major consumer of many important materials. For example, around a third of petroleum is consumed by passenger cars and trucks in the form of gasoline and diesel fuel. Approximately 10% of all plastics, which also come from oil, and 15% of the world's steel are consumed by the auto industry. Nearly two-thirds of all natural and synthetic rubber goes into making automotive tires and components.

Sustainable approaches to automotive manufacturing must start with designing vehicles that use less material and energy to make, as well as designing vehicles to last longer and be easier to recycle. Although automobiles are one of the most recycled products, as shown in **Figure 9.2**, there is still a significant amount of waste generated annually from automobile manufacturing as well as from end-of-life vehicles that are not recycled, though precise global figures are difficult to find.
In addition to the vehicles themselves it is also estimated that for every

vehicle produced, just over a quarter of the total material input becomes manufacturing waste. For steel, this means that over 15 million tons of scrap metal is generated each year by the global auto industry.

Figure 9.2 Status of automotive recycling.

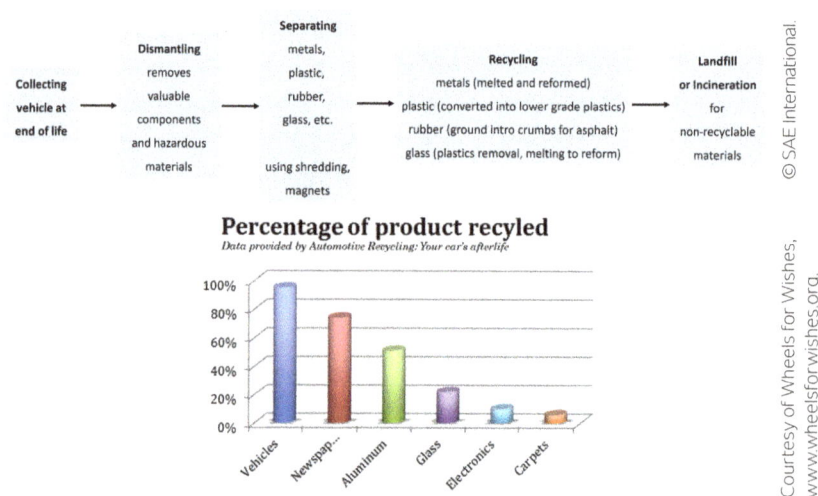

EVs are even heavier than ICEVs, but they will provide additional metal recycling opportunities due to the high value of materials used in them. Electric motors, for example, use steel, copper, and rare earth magnets, and can be refurbished along with the power electronics, BMS, and wiring harnesses. Large battery packs contain hundreds of pounds of valuable raw materials, such as lithium, cobalt, and nickel, and can either be recycled or repurposed for secondary "second-life" energy storage applications at the end of their automotive service life. However, even with efficient battery manufacturing processes, there is still a 5–10% scrap rate on average [9.1]. Improving the collection rate and the recycling infrastructure will be necessary for recycling to become economically viable because, at present, some materials are inevitably lost or degraded, while some battery applications require higher-purity materials than can be achieved through recycling alone. There can also be technological limitations in efficiently recovering certain materials. However, even if recycling cannot meet all the short-term needs for battery EV materials, it can still significantly reduce the amounts needed to be mined, which will reduce the associated environmental burden and boost supply chain security. In short, a mix of strategies will be needed, including better battery collection systems and advanced recycling, designing batteries for easier disassembly, reusing and remanufacturing to extend battery life, and developing alternative battery chemistries that use more abundant and easily recyclable materials.

Automotive plastics and composites, typically used because they are lightweight, are some of the most challenging materials to recycle economically because the value of these mixed materials rarely justifies the cost of labor to disassemble and separate them for recycling. However, using commonly available recycled materials like plastic bottles and cardboard could be more feasible for the type of low-speed urban EVs described in Chapter 8 than it is for conventional cars for two main reasons. For a start, the lower speed ratings (typically under 25 mph) and safety standards mean that the recycled plastic/fiber composites do not need to have such high strength or impact resistance. Second, the lower volume production runs are likely to be more feasible with recycled materials than with traditional automobiles where the upfront costs of new tooling for recycled parts could be prohibitive. Recycled plastics are generally cheaper than the virgin resins needed for automotive-grade plastics, and if the value of reducing landfill waste is factored in by the municipality, then the economics can become even more favorable.

In the context of an urban mobility strategy where low-speed EVs are designed, developed, and produced locally to complement the public transport system and to provide last-mile connectivity, it is not a stretch to imagine that the municipality could put a value on not only the jobs being created but also the landfill being reduced (and this will be discussed further in Section 9.5). This could be a variation on the concept of industrial symbiosis, where one industry's waste becomes another's raw material, and it is gaining traction in developing countries. In India, for example, cement plants are using waste materials, such as fly ash and slag, from other industries as alternative raw materials, reducing resource consumption and waste. Sourcing raw materials locally also strengthens supply resilience and significantly reduces the emissions embodied from shipping parts and commodities long distances around the world.

The low weight of recycled plastic/fiber composites makes the EV lighter and helps to reduce the size of the battery needed, and this makes the vehicle even lighter, cheaper, and more efficient. Many low-speed vehicle manufacturers already use recycled plastics and other sustainable materials in body panels, interior components, and even structural pieces because of these cost, weight, and performance advantages. In short, the lower technical demands make these more affordable recycling options quite viable.

Plastic water bottles, for example, are typically made from polyethylene terephthalate (PET) or high-density polyethylene (HDPE). These thermoplastics can be shredded, melted, and remolded to create plastic resins suitable for making interior car parts like underfloor paneling, the trunk liner, and other interior trim. It can also be used in pellet form to make filler material for thermal insulation and acoustic isolation. Cardboard can be recycled into a paper fiber that can be mixed with the plastic resin to make interior parts for the vehicle (headliners, seat cushions, etc.). When reinforced with recycled glass fibers, the plastic can be used to make stronger and more durable composite materials for other vehicle parts, such as headlight and taillight housings, bumpers, pillar reinforcements, and seat frames. Recycled rubber from used tires can be remade into new tires and floor mats.

Major automakers like Ford, Toyota, and BMW are already incorporating recycled plastic bottles and cardboard into various interior and nonstructural car parts and components. Mercedes-Benz is developing a synthetic leather that can be made from a combination of used tires, agricultural waste, and bio-based polymers. It can use re-tanning processes to produce the same look, feel, and aging as natural leather but with a reduced environmental footprint [9.2]. Although more research is needed to optimize strength for broader usage in the auto industry, there should be many promising opportunities for using these recycled waste materials in low-speed EVs. This will lessen the need to make virgin plastic resins from petroleum and will reduce landfill waste at the same time, giving these waste materials another life in vehicle parts production. After all, if PET can be recycled up to ten times, it could last many more years if it is used in a car instead of in a frequently recycled water bottle! Mercedes-Benz is developing a synthetic leather that can be made from a combination of used tires, agricultural waste, and bio-based polymers. It can use re-tanning processes to produce the same look, feel, and aging as natural leather but with a reduced environmental footprint [9.2].

Can more be done to improve recycling efficiency? Perhaps lessons can be learned from the Zabbaleen, a group of "garbage men" in Cairo, who go door-to-door and collect two-thirds of all the garbage discarded by more than 20 million Cairenes and recycle up to 80% of it. This is a rate three times higher than in most US cities. Their work has received attention from Foundations around the world which has helped to fund acquisition of plastic crushers and cloth shredders [9.3]. Emulating this level of recycling that is being achieved in Cairo could significantly reduce pollution and energy consumption while creating employment and lowering vehicle material cost (**Figure 9.3**).

Figure 9.3 Extensive recycling in Cairo.

9.3. Natural Materials

Nature itself provides abundant sources of lightweight, renewable materials that could be alternatives to conventional, petroleum-based plastics and composites, especially if we rethink design constraints on the vehicle. Natural, bio-based materials consume carbon dioxide from the atmosphere during their growth, which offsets the environmental impact from processing them into final parts. These materials can also be recycled, of course, and could be well-suited for making low-speed, last-mile-type vehicle components where the requirements may be less challenging than with automobile production. Many have promising applications, particularly in the developing world where agricultural feedstock sits close to urban centers and where labor is cheap and manufacturing processes that use locally sourced materials could be easier to implement, creating employment opportunities.

Early cars used natural materials extensively as they incorporated significant amounts of wood in their construction, particularly in their body frames (known as "woodie" cars). This was because horse carriage makers transitioned into early car manufacturing and had widely available woodcraft skills and tools. Wood was abundant and easy to work with and repair, and it is flexible with a good strength-to-weight ratio (**Figure 9.4**). However, wood is no longer commonly used to make cars because it provides less crash protection than modern materials and can splinter dangerously. It also has inconsistent structural properties and is heavier than modern composites. Wood is also susceptible to rot and weather damage but presumably these concerns are being addressed by Dacora, the first automaker to be founded by a woman. Its first vehicle, a luxury EV, uses handcrafted wood prominently for the hood and dashboard as a way to differentiate itself in the marketplace.

Figure 9.4 Wood was a significant automotive material at one time.

Daderot/https://commons.wikimedia.org/wiki/File:1939_Chevrolet_woodie_station_wagon_-_Automobile_Driving_Museum_-_El_Segundo,_CA_-_DSC01656.jpg.

Daderot/https://commons.wikimedia.org/wiki/File:1940_Chevrolet_Special_Deluxe_woodie_station_wagon_-_Automobile_Driving_Museum_-_El_Segundo,_CA_-_DSC01692.jpg.

For lightweight, low-speed EVs operating in dry climates, wood may have an opportunity, but bamboo shows even more promise. Compared to wood, bamboo grows much faster and is stronger, more flexible, and lightweight because of its hollow nature. It also has more consistent properties and is not affected as much by moisture (**Figure 9.5**).

Figure 9.5 Bamboo, flax, and hemp are promising materials for vehicle construction, particularly for low-speed vehicles.

Zambia-based Zambikes already makes bicycle frames from bamboo, and the approach could be extended to low-speed EV frames and body panels. Jute and sisal, which also grow in the tropics, are well-suited for making seats and insulation, while rice husks, coconut husks, and wheat straw can all be used to reinforce composites or as filler materials for nonstructural components. Soy-based resins or foams can be made into cushioning, insulation, and interior components.

Hemp fiber is extremely versatile and can be used to make textiles, construction materials, biodegradable plastics, and reinforced composite materials. These composites can be used to manufacture body panels, interior trim, and nonstructural parts that are light and strong. Hemp fibers also have good thermal and acoustic insulation properties, making them suitable for insulating material, and they can be woven or blended with other materials to create fabrics for seats, door panels, and other interior applications. Automakers including BMW, Ford, and Mercedes-Benz have experimented with using hemp-based components in vehicles for door panels, trunk liners, and interior trim because they can be lighter and more environmentally sustainable than petroleum-based materials.

In some developed markets, where there is a market for medicinal and recreational marijuana, an urban marijuana farm could produce hemp as a by-product because hemp and marijuana are different varieties of the same plant species, *Cannabis sativa*. Instead of being rejected as waste because it interferes with the female plant's harvesting into marijuana, the remaining male plant stalks could be used to obtain industrial hemp from the outer bark layer and its inner core. The outer bark contains long hemp fibers, which are stronger and lighter than glass fibers, and it can be processed to create hemp fiber-reinforced plastics for interior panels and trunk liners. The inner core contains lightweight and porous woody hemp hurds, which can be used to create insulation materials or to replace clays as a sustainable filler. This could create another form of industrial symbiosis, whereby the male plants, otherwise considered waste on a marijuana farm, could feasibly provide a hemp by-product stream usable for low-speed EV production nearby in the same city, assuming it is properly separated and regulated.

While many of these materials can also be used in conventional automobiles, they are far better suited to low-speed, last-mile vehicles that can accept less strength and durability, and where there is more emphasis on lightweight construction to improve energy efficiency and maneuverability, and perhaps a greater willingness to experiment. As with recycled materials, the manufacturing approach used for low-speed vehicles needs to be cost-effective and suitable for small-scale or localized production, and this could work in favor of these materials that do not require such precise, durable, powerful, and costly tooling as is needed for conventional automobiles.

The suitability of these materials for vehicle components will depend, clearly, on factors such as their mechanical properties, durability, manufacturing processes, and compliance with any safety and regulatory standards, so further research will be needed to prove their viability, even for low-speed EVs. However, it is quite possible that the major challenge to the adoption of recycled and natural materials will be in collecting and processing them at a cost that makes them competitive with conventional, mass-produced, and petroleum-based plastics.

In summary, with less stringent vehicle performance requirements for low-speed EVs, there is potential to exploit "unconventional" automotive materials such as those recovered from local landfill waste, which includes plastics and even cardboard, as well as agricultural waste, such as hemp and bamboo. Of course, aluminum and other lightweight metals can also be recycled from local waste and can be used to make vehicle structures and parts. In this way, vehicle production as well as vehicle usage can be made more sustainable and, potentially, more affordable.

9.4.
Localized Production

Traditional automotive factories or "plants" can easily cost $1B to build, and some may cost as much as $5B. This spread is due to several factors, including the cost to acquire land, to set up the infrastructure (for connecting to the electric grid, transport network, and water supplies), and to build the factory to meet environmental, safety, and regulatory standards. It also depends on the level of automation and whether the plant also makes sub-assemblies, like stamping of door panels, in addition to the final vehicle assembly. BYD is taking vertical integration to such levels that it is building a factory in Zhengzhou, China, that is 32,000 acres (50 square miles or larger than San Francisco!) [9.4]. Finally, it also depends on how many vehicles the automaker plans to make each year in the plant and how much flexibility it wants to build to adjust to changing corporate, regulatory, or market conditions.

An alternative approach is likely to be more appropriate for producing relatively low volumes of low-speed EVs, perhaps in a downtown area, while still remaining agile to changing customer demands. This approach is based on the microfactory (**Figure 9.6**). While traditional automobile plants may occupy more than a square mile, operate 24/7, and require just-in-time deliveries needing many large trucks arriving and departing each day, micro-factories can be located downtown. By minimizing the huge capital costs of building a typical automotive manufacturing plant, the initial upfront cost for the microfactory can be at least ten times less or below $100M. For most businesses attempting to disrupt the status quo, this is a critical benefit in managing risk.

Figure 9.6 Comparison of a traditional (greenfield) auto manufacturing plant and a (downtown) auto microfactory.

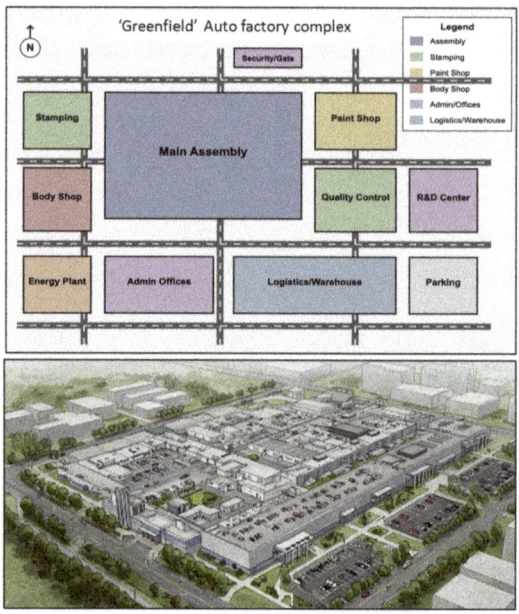

Approximate land area: 200 acres (80 hectares)
Approximate number of employees: 5,000
Approximate number of vehicles produced per year: 500,000

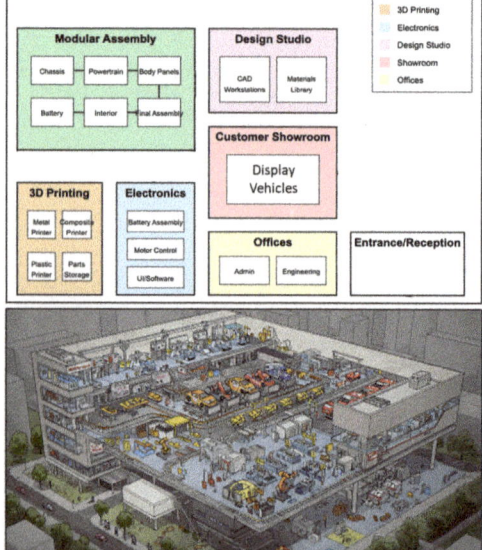

Approximate land area: 1 acre (0.3 hectares)
Approximate number of employees: 100
Approximate number of vehicles produced per year: 10,000

© SAE International.

A microfactory differs in several other important ways from a typical large-scale auto factory in terms of flexibility, automation, localization, and supply chain. Microfactories are small-scale facilities, often targeting production of 10,000 vehicles/year. Hyundai has opened a microfactory in densely populated Singapore that can produce 30,000 EVs/year. Typical auto plants, however, can easily produce ten times this volume and some can even produce more than one million vehicles/year. Microfactories are designed for rapid changeovers and the ability to quickly adapt to changing market demands, unlike traditional auto plants that are highly optimized for high-volume, standardized production runs. Perhaps surprisingly, microfactories rely even more on automation because they need to be agile and quicker to incorporate the latest robotics and AI advances. Automation also reduces labor costs and increases production speed, which is why microfactories are built from the ground up with automation in mind. While traditional auto plants also use automation, their larger scale and more established processes and labor policies sometimes limit the extent of automation.

Microfactories tend to focus on serving local or regional markets, tailoring products to local tastes. Conventional auto plants, on the other hand, produce vehicles for regional or even global distribution. The local microfactory approach dovetails well with using the type of locally available materials, either recycled or natural, described earlier in this chapter. Although all vehicle manufacturing involves integrating parts from suppliers, the microfactory approach takes this reliance even further because it eliminates costly stamping and painting processes, focuses more on final vehicle assembly, and relies on nearly all the components and modules to be pre-engineered and shipped in by suppliers. This reliance on suppliers could be extended to automakers if, for example, an automaker supplied or licensed the ACE platform, described in Chapter 8, Section 8.5. It could, for example, achieve economies of scale and supply a version of it to microfactories, which would then focus on making just the coach and mating it to the ACE platform. By simplifying the manufacturing process, there is far more opportunity for a microfactory to support do-it-yourself (DIY) or local community-based assembly.

The advantages of a smaller, more flexible manufacturing process are particularly well-suited for low-speed, geofenced vehicles that will be used for local delivery vans, campus shuttles, or robopod fleets because vehicles confined to limited geographic areas or specific use cases should have lower overall production volumes compared to high-volume consumer passenger cars, at least in the near term, and this makes the economics of a more flexible manufacturing process more viable. Low-speed, geofenced vehicles are designed to be compact and lightweight for densely populated environments, and recycled plastic or natural material composites and microfactory assembly techniques may help achieve these size and weight objectives better than using conventional metals and traditional heavy stamping processes. These vehicles may not need the very high strength and rigidity of stamped metal body panels, so composites that are lighter in weight may suffice. With more predictable route, terrain, climate, speed, and range requirements, and with more emphasis on fleet sales instead of consumer sales, the automation is less complex and the paint shop may not even be needed, saving cost versus mainstream passenger vehicles that are intended for general highway use, "overengineered" and even, perhaps, aspirational or emotional purchases.

Microfactories enable closer proximity to local material inputs and the intended customer base so that a quicker response can be made to demand changes. Low-speed EVs will likely be designed with a modular vehicle architecture using preassembled key components from suppliers that leverage their own expertise and economies of scale. A faster, simpler final assembly process can then occur at the microfactory. The vehicles can be designed with swappable, upgradeable components and accommodate new technologies and customer needs over time, extending the useful vehicle life with hardware as well as software upgrades.

Although most components will be sourced from suppliers, there could be opportunities to produce parts cheaply on-site in the microfactory by leveraging 3D printing for select low-volume, customized parts. Rather than traditional manufacturing methods that remove material, like machining, additive processes like 3D printing build objects layer by layer and are able to create very complex geometries that are optimized for strength, weight, and functionality. This reduces material waste compared to machining processes, and as 3D printing scales up, it should enable on-demand production of components locally with minimal inventory and transport needs. 3D printing technology has been adopted by the automotive industry in recent years to develop full-scale prototypes of cars, interior components (e.g., dashboard, door panels, center consoles), and exterior components (e.g., side mirrors, spoilers, grilles) for rapid testing purposes. In the manufacturing arena, it can be a quick and inexpensive way to build fixtures for assembly processes and molds for low-volume production runs while in the aftermarket space, spare parts have been made for customized accessories, such as gearshift knobs, as well as for discontinued parts (**Table 9.1**).

Table 9.1 Comparison of various automotive manufacturing processes.

	3D printing (additive manufacturing)	Injection molding	CNC machining	Casting/forging
Process	Layer-by-layer material deposition	Molten material forced into a mold	Material removed via cutting tools	Molten metal poured into a mold (casting) or shaped under pressure (forging)
Lead time	Short (no tooling required)	Long (mold creation needed)	Moderate (programming and setup)	Long (mold/die creation)
Cost (low volume)	Low (no tooling)	High (mold costs)	Moderate (machine time)	High (mold/die costs)
Cost (high volume)	High (slow per-part cost)	Low (economies of scale)	High (labor and time)	Low (bulk production)
Design flexibility	Very high (complex geometries possible)	Limited by mold design	Moderate (constrained by tool access)	Limited by mold/die
Material options	Plastics, metals, composites, ceramics	Mostly plastics and some metals	Metals, plastics, wood	Mostly metals
Surface finish	Rough (often requires post-processing)	Smooth (mold finish)	Smooth (machined)	Moderate (may need finishing)
Waste	Minimal (only uses needed material)	High (runners, sprues)	High (material removal)	Moderate (excess material)
Best for	Prototypes, custom parts, complex designs	Mass production of identical parts	Precision parts	High-strength structural parts

© SAE International.

While 3D printing is currently used for prototyping, tooling, and low-volume production of specialized parts, it is not yet an economical method for the high run rates needed for mass production. As the technology advances, perhaps more automotive components could be 3D-printed, but far more opportunities exist to leverage 3D printing for low-speed EVs that may not have to meet all the traditional automotive crash safety standards. Combining 3D-printed fibers and a plastic resin inside an inexpensive soft mold, which is more economical for lower production quantities, might even make it possible to form larger and more complex shapes such as hoods and doors.

The main consumer benefit of 3D printing is customization, but other benefits, perhaps more important to the manufacturer, include rapidly prototyping and testing parts, reducing weight by eliminating "unnecessary" material, reducing inventory because parts can be printed on demand, and reducing tooling costs. In conclusion, lower microfactory production volumes with less stringent performance requirements for parts may end up being a very good fit for 3D printing. Although several companies, such as Local Motors (**Figure 9.7**), and Arrival have been unsuccessful at bringing a 3D-printed vehicle to market, the YOYO XEV, made by an Italian-Hong Kong startup, is the world's first production vehicle to extensively use 3D printing and is available in Italy and other European markets with a base price of 9900 euros (**Figure 9.8**). There may still be regulatory and consumer safety-related concerns to adoption, but both of these can, perhaps, be mitigated with fleet vehicles operating in a low-speed environment, particularly in "car-free" or "car speed-controlled" zones.

Figure 9.7 The author in the driver's seat of a Local Motors' Strati, with a 3D printed body, and Local Motors' Olli roboshuttles.

Figure 9.8 YOYO XEV use 3D-printing for body panels and the instrument panel.

From a sustainability perspective, microfactories are well-suited to producing low-speed EVs at a lower cost than a traditional auto plant, and they can reduce the energy and emissions associated with vehicle production. They may also be able to leverage not only locally available materials, like landfill waste, but also locally available energy sources, such as landfill gas. Setting up manufacturing facilities closer to end markets reduces the need for long-distance shipping, further reducing energy consumption and emissions and fostering local economies.

In short, transitioning to a more affordable and sustainable mobility model requires rethinking not only the vehicle design and performance, as described in Chapter 8, but also the conventional large-scale, centralized manufacturing approach. A new distributed manufacturing concept can enable this transition from mega-scale auto plants, requiring massive capital, to a distributed network of smaller "microfactories" localized in each community so that they can manufacture the type of vehicles they need for local usage to provide "last-mile" connectivity with public transport and for local delivery of goods. These compact facilities have lower costs by focusing on specific vehicle models tailored to regional needs. With a distributed regional network, shorter supply chains and sourcing of local materials and labor, the risk of a disruption to a complex global supply chain can also be reduced.

With smaller production volumes, microfactories can offer greater customization for vehicles and can adapt quickly to changing market demands and product iterations because of their proximity to the customer base, smaller scale, and digital manufacturing processes.

Compared to traditional factories, microfactories require a lower initial investment in infrastructure and machinery, and several microfactories can be established in different regions, further reducing the risk of supply chain disruptions.

The downside to this rosy picture, and what has prevented it from catching on so far, is that microfactories have a lower overall output capacity compared to large-scale traditional factories and this usually results in higher unit costs because of fewer economies of scale, a less established supply chain, and less proven manufacturing processes. However, if the ACE platform is sourced from a mass production supplier or automaker, then economies of scale can be achieved for the more technically challenging platform while still allowing a local manufacturer to build and customize the coach in the microfactory and to develop software for the customer and fleet application. Moreover, many cities and countries face the same basic challenge of moving people safely and cost-effectively on congested roads, so it is likely that a vehicle developed in a microfactory in one location could export their excess micro-EV production to neighboring cities or regions to generate greater scale economies.

By situating the microfactories alongside facilities producing sustainable material inputs such as recycled plastics and natural fibers (hemp, bamboo, etc.), a new form of local "vertical integration" may be possible with raw material collection and production coupled directly to component and vehicle manufacturing in one centralized zone. This maximizes resource efficiency and circularity, and by having multiple stages of production colocated, the need for transporting components or materials between facilities is reduced, leading to shorter lead times and lower logistics costs. Coordination between teams and departments becomes easier, enabling better collaboration and tighter control over the entire production process, which makes it easier to identify and address quality issues promptly. Having vertically integrated operations within a single location enables faster adjustments and adaptation to the changing market demands or production requirements. There are several trade-offs, however, as a more vertically integrated facility increases the upfront capital investment required for land, buildings, and equipment. Moreover, although greater control of the supply chain is achieved with vertical integration, geographic concentration can increase vulnerability to disruptions caused by natural disasters, for example.

The decision to pursue colocated verticalization depends on factors such as the nature of the product, the complexity of the supply chain, the need for agility and responsiveness, and the availability of resources and capital. The viability of the microfactory approach will probably require a holistic viewpoint that focuses on circularity where waste from a local business can be turned into an affordable feedstock and where the right balance is struck between standardization using a common ACE platform and customization of the coach. It also relies on collaborating closely with local government, businesses, and end users to ensure supportive policies are in place for cost-effective materials supply, waste management, production, employment, and transport policy integration as will be discussed in Section 9.5. If these considerations can be met, then there is a good chance for success and scaling to other locations.

A variation on this local assembly can also be applied to the e-kit, described in Chapter 8. As an example, in rural, developing markets the e-kit might be assembled in a major city or port, such as Nairobi or Lagos, where there are the required commercial, engineering, and production skills and easier access to componentry, such as imported batteries, motors, and solar panels. The e-kit could then be distributed and sold to rural, agricultural customers inside the country or even exported to nearby countries. A leasing model may be required to address the prohibitive upfront cost for many subsistence farmers, or the financing might copy M-Kopa's pay-as-you-go solar energy model that is designed to make solar LED home systems affordable in East Africa. In this model, that was shown in **Figure 8.14**, the solar system includes a SIM card so that if payments are missed then the system can be remotely disabled. The aim is to support mobile micropayments, which would otherwise be made for kerosene-fueled lighting, to the point where more than five million customers own the system outright [9.5].

9.5.
Economic Benefits and Supportive Government Policies

In addition to supporting local production, cities have a potential mix of incentives and deterrents at their disposal to influence how people and goods move around inside their perimeter. However, public support will still be essential to implement and enforce any new transport measure.

For example, there are several reasons why London was able to implement congestion charging in 2003 but it could only get started in 2024 in New York, a US city with similar congestion issues and public transport system service. One reason is that London's central city area is more compact and well-defined, making it easier to implement a cordon-based congestion charge system covering approximately eight square miles. Congestion charging has been publicly accepted in London, where residents have seen the benefits of reduced traffic and air pollution. The congestion charge did raise traffic speeds at first but over time the reallocation of space for bicycles as well as increased goods delivery and ride-hailing has meant that traffic speeds are now back to around the same level as before the congestion charge was implemented, but Transport for London (TfL) was able to use the congestion charge revenues to invest in public transport improvements as promised. Over $300M was raised by the congestion charge in 2023, for example, and was reinvested in improving the bus and subway network, road and bridge maintenance, and in cycling infrastructure [9.6].

Moreover, London has an integrated transport organization, TfL, that not only manages the public transport network but also regulates taxis, manages the road system, plans future transport infrastructure developments, and actively supports walking and cycling modes. TfL is the key agency responsible for planning, managing, and improving all aspects of transportation within Greater London's 607 square mile area, and its wide-ranging remit is crucial for coordinating the city's complex mobility needs. There is no equivalent in any US city, and New York, for example, has struggled for many years to bring together the necessary state and local agencies to do the same even though its Metropolitan Transit Authority believes congestion pricing would cut traffic by 17%, improve air quality, and increase public transport use by 1–2% [9.7]. It has only recently been introduced but early indications include one million fewer vehicles entering the New York City congestion charge zone with fewer road accidents, while increasing subway ridership 7.3% on weekdays. Nearly $50M has been raised by the toll, and it is expected that $15B can be raised over time to fund capital improvements for public transport [9.8]. If traffic speeds can be increased through congestion charging, then ride-hailing and goods delivery businesses could make productivity gains that more than offset the charge [9.8], and as robotaxis and automated vans become more prevalent on dense, urban streets, there could even be "road space auction" deals made between these fleet operators that are a win–win for the city and for business!

Some European cities have also created low-emission zones that tax ICEVs because of air pollution concerns (**Figure 9.9**). For example, London expanded its LEZ in 2023 and measured a 31% decrease in small particulate emissions (PM2.5) during 2024 [9.9]. This is separate from the congestion charge that applies to most personal ICEVs but has an EV exemption. Other cities are going further and looking to establish vehicle-restricted zones, residential parking permit areas, or even outright bans on personal vehicles from accessing city centers during certain times, or even permanently. Central Paris's ZTL (limited traffic zone) is an example of this. These policies are driven by environmental concerns around air pollution and noise levels, as well as trying to free up parking space and reduce traffic congestion to make the city more "livable" by encouraging active modes of transport (walking, cycling) and public transport. However, it also acts as a powerful enabler for fundamentally rethinking automotive solutions that are not bound by the traditional design constraints and requirements that are optimized for high-speed highway operation, such as microEVs and robopods. Helsinki has set a goal of becoming a car-free city by 2050 by investing in green spaces, pedestrianized areas, cycling lanes, bike sharing, and MaaS [9.10].

Figure 9.9 A global sample of cities putting a curb on personal vehicle use.

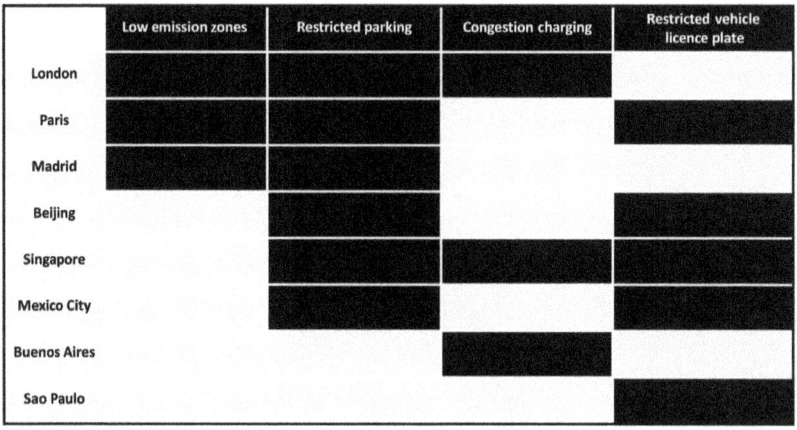

	Low emission zones	Restricted parking	Congestion charging	Restricted vehicle licence plate
London	■	■	■	
Paris	■			■
Madrid	■	■		
Beijing		■		■
Singapore		■	■	■
Mexico City		■		■
Buenos Aires			■	
Sao Paulo				■

© SAE International.

To further gain public support and understanding, disabled drivers and passengers, public transport (taxis and buses), emergency services (fire, ambulance, police), and EVs were all exempt from paying the London congestion charge, and, importantly, residents living within the congestion charge area can apply for a 90% discount. Goods delivery vehicles need to pay these charges unless they are EVs or deliver in the evening or early morning when congestion is not usually such an issue. Similar exemptions apply in those city zones that restrict vehicle access or have a limited traffic zone, such as Paris' ZTL and Barcelona's Superblock urban design, shown in **Figure 7.2**. The aim is to ensure a sweet spot is reached whereby residents generally understand and approve of the restrictions, can live with them, and believe the benefits outweigh the drawbacks. Unlike London's congestion charge, these zones do not generate revenue directly from the policy, but their aim is to improve air quality, public safety, and overall quality of life. From these "green shoot" restricted traffic zones, it should be easier to add electronic vehicle speed control, as described in Chapter 5, Section 5.5.2, to ensure that no vehicles in the zone are traveling above the local speed limit because they will either be owned by residents or by businesses (delivery, taxi, and ride-hailing) and local government (fire, ambulance, police) that support the restrictions. This development would make it easier to introduce geo-fenced micro-EVs or robopods to provide last-mile connectivity to public transport stations inside the speed-controlled zone because users would feel safer using them (**Figure 9.10**).

Figure 9.10 Hague's car-free city center.

Even without a congestion charge, there can be significant economic gains for restricted traffic zones if they go hand in hand with local production of micro-EVs and robopods. To encourage municipal or regional governments to support a rethinking of the transport system along the lines discussed in the last few chapters, it will be necessary to clearly lay out these benefits, which include more affordable and accessible transport, new employment opportunities, waste reduction, improved air quality, and road safety with more walkable streets, plus an intangible, which is increased civic pride and engagement. Quantifying these outcomes will be important to overcome the many challenges and opposition that face this new paradigm. The following is intended merely to illustrate the possible magnitude of some different types of value streams that may help the business case for cities to support such a change.

Reducing plastic waste in landfills, for example, can lower waste management expenses because municipal solid waste landfill costs are roughly $70/ton of waste, and recycling plastics into vehicles could divert thousands of tons of waste annually. Reduced landfill waste also helps cities meet sustainability targets and avoid future environmental penalties. For the EV manufacturer, recycled plastic is likely to be cheaper than virgin plastic, and this will lower the material procurement expenses. There may also be tax incentives for using greener materials. Lower vehicle production costs, reduced waste management expenses, a positive brand image that can attract potential investment, and the creation of local recycling and manufacturing jobs are some of the longer-term benefits. For a typical microfactory that might produce 10,000 EVs annually, the waste management savings could be up to $1M/year while a 10% savings on vehicle manufacturing cost using lower-cost recycled materials could equate to $2M/year, assuming materials for the vehicle body and frame account for approximately one-third of the vehicle's manufacturing cost or around one-fifth of the vehicle cost, assumed to be $10,000.

In terms of employment, there may be 100 high-quality manufacturing jobs per microfactory, and with an average annual wage of $60,000 per worker, the total payroll might come to $6M. A rough rule of thumb is that each manufacturing job supports two indirect jobs so an estimated 200 additional jobs might be needed in raw material processing, supply chain logistics, maintenance and engineering, and local services. The creation of each direct manufacturing job typically generates about twice as much additional economic activity, meaning that a $6M payroll adds approximately $12M to the local economy. Taxes on income, property, and local sales also increase economic activity and provide additional tax revenues of around $1M annually. Finally, high-tech manufacturing creates opportunities for technical training programs and partnerships with local colleges. Conservatively, the total economic benefit generated by the microfactory could reach $15M annually, plus long-term sustainable infrastructure development.

After reduced landfill costs and local employment generation, the third benefit to the city is in public transport. In New York City, there are roughly 1.6 billion passenger journeys per year on the subway and bus system run by its Metropolitan Transportation Authority (MTA) [9.11] and the subsidy is $7.2B in 2023 [9.12], humming to the tune of around $4.50 per passenger trip. Even with this subsidy, fewer than 15% of the city's population uses public transport mainly because of the first- and last-mile transport gaps. However, if robopods could connect people to public transport more easily, this would boost the ridership, slashing the subsidy per rider as well as reducing congestion. The current annual public transport subsidy for a mid-sized city (one million people) is typically around $50M, which means that if ridership doubles then the annual savings could be very significant. With fewer people driving in cars, there could be less need for road maintenance and parking infrastructure. The economic value from reduced congestion and fewer carbon emissions, as well as improved public transport that provides more accessibility for the elderly and disabled, are some of the indirect benefits. Cumulatively, all of these various economic benefits to the city might exceed $50M/year for a mid-sized city and substantially more for a megacity.

The economics of robopod operation could be even more favorable than for robotaxis because they should have lower purchase and operating cost. Because they are considerably smaller than robotaxis, the real estate cost for parking them can be less, and as they will be serving public transport and providing more affordable accessibility, they could be treated more favorably by the city and not face some of the penalties associated with "zombie" operation or potential pickup and drop-off curb fees.

How might cities support the rollout of this new "circular" mobility ecosystem that transforms waste into vehicles, creates local employment and improves accessibility and mobility for everyone? Some cities could begin with their existing restricted traffic zones, enforcing speed limits and laying the foundation for introducing low-speed micro-EVs and robopods that will move people and goods. This will require defining the zone with appropriate urban core and school zone integration and possibly with some time-based and special event management variations. In addition, they could create local manufacturing incentive programs with tax incentives for local production and waive the local sales tax on machinery and tool purchases. The city can update zoning policies to specifically allow local manufacturing, with the appropriate noise, operating hours, and environmental and safety standards, and the manufacturer might receive local production credits, innovation grants, and circular economy and job creation benefits.

The city can also share its existing mobility data to help guide pilot program parameters and metrics, and support pilot programs for demonstrating how these vehicles can integrate with public transport to improve accessibility and reduce congestion. Such integration may involve constructing new charging spots at public transport station hubs and possibly adding some smart traffic systems as well as real-time information and unified payment systems. These low-speed, geo-fenced, and speed-controlled micro-EVs might even share lanes currently dedicated to bicycles and buses to enhance their safety and traffic speed. The purpose of a pilot program, which might initially use existing micro-EVs, would be to test operational effectiveness, gather consumer feedback, and assess the economic costs and benefits. If the outcome is viewed positively, then there could be a gradual expansion of the operating envelope (area coverage and time windows, such as night-time deliveries to reduce congestion during business hours) and the introduction of microfactory-built micro-EVs, starting with simple assembly operations and then, over time, increasing local content using recycled and natural materials.

The micro-EVs could evolve into robopods, with a high degree of V2X connectivity, for high-reliability deployment and taking lessons from robotaxi development in terms of dynamic rebalancing, automated charging and maintenance, and usage optimization. One of the advantages of robopods is that their autonomous operation relaxes the need to install parking space at public transport stations and, instead, they can park in nearby existing parking spots and be summoned by riders on demand. They could even be recharged there if there is a partnership between the MaaS and parking operators. The leveraging of existing assets can help to avoid the disruption and costs of installing new parking and charging infrastructure.

The success of such a program will be determined by a variety of factors, such as clear cost advantages, reduced traffic congestion and environmental pollution, a strong safety record, and community support for both an expanded MaaS and urban policies that support low-speed zones and alternative transportation (**Figure 9.11**).

Figure 9.11 A local economic mobility system that uses locally available and renewable materials to build low speed EVs that can be used to support public transport with last-mile connectivity.

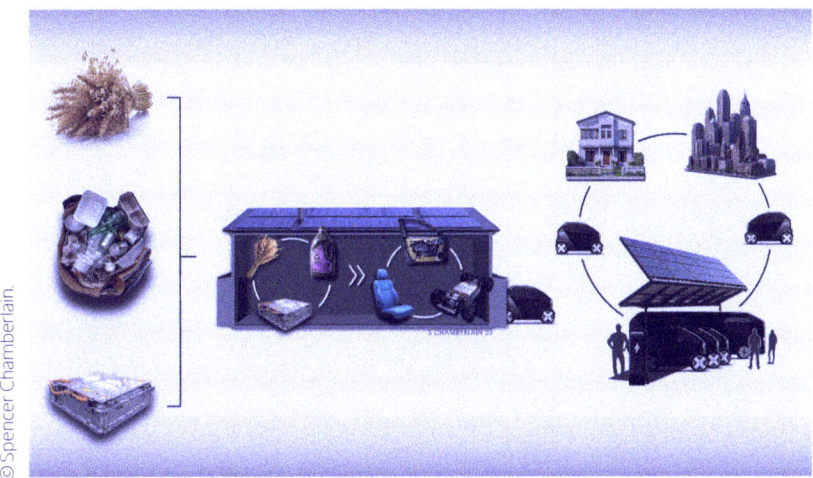

In addition to local government policies, there are additional measures that can be enacted at the regional or national level, many of which have been executed by the Chinese government in their support of EV production and market adoption. These include tax credits or deductions for companies that manufacture products locally and direct financial support to local manufacturers through subsidies, grants, or low-interest loans to help offset the costs of setting up new facilities, upgrading equipment, or training workers. Electricity rates can be discounted for manufacturers to reduce production costs, and R&D grants can help local manufacturers innovate and develop new products or processes.

Governments can also give preference to locally manufactured goods when making procurement decisions. Cities, after all, are increasingly incorporating EVs into their own municipal vehicle fleets, creating a steady demand for urban-focused EV models and providing a guaranteed market for automakers to develop and manufacture suitable EV models for city use. The city can fund educational and vocational training programs to develop a skilled workforce for local manufacturing industries, and they can invest in roads, ports, and railways, to facilitate the movement of raw materials and finished goods for local manufacturers. Manufacturing clusters or hubs can be encouraged to create economies of scale and foster collaboration among local manufacturers, like EV City in Zhaoqing (China) and the EV Industrial Park in Pune (India). Cities can be effective in bringing together automakers, tech companies, energy providers, and other stakeholders to codevelop EVs and the associated EV ecosystem. These partnerships can enable the coordination of vehicle manufacturing, charging infrastructure, renewable energy, broadband connectivity, and smart city initiatives. Regulations, permitting processes, and zoning laws can be streamlined to make it easier for manufacturers to establish or expand operations locally.

While governments often provide subsidies and incentives for EVs to promote their adoption and reduce emissions, it is worth noting that micromobility solutions like micro-EVs and e-bikes have typically not received the same level of subsidies. EVs may be viewed as a more significant contributor to reducing greenhouse gas emissions and air pollution compared to e-bikes and scooters, or perhaps e-bikes and scooters are seen as being more affordable, which reduces the need for direct subsidies compared to the higher upfront costs of EVs. The regulation of micromobility solutions, particularly shared services, is still evolving, and governments may be hesitant to provide subsidies until clear regulations and safety standards are established. Another explanation is that EVs require substantial investments in charging infrastructure and grid upgrades, which governments may prioritize over micromobility infrastructure. Perhaps the main reason, though, is that the automotive industry is a significant economic driver, and governments may prioritize incentives for EV adoption to support domestic manufacturers and job creation. Even without direct subsidies for micromobility solutions, governments can still support their adoption through investments in dedicated infrastructure, such as bike lanes and parking facilities, and in setting a regulatory framework that encourages their safe and responsible use.

In summary, priorities should shift from conventional high-margin premium vehicles developed and marketed by multinational conglomerates toward a focus on democratized mobility and accessibility that is tailored cost-effectively to communities' needs, providing local job opportunities in product development, manufacturing, and supply chains. Overall, this decentralized, regionalized model has many potential societal benefits as it can provide affordable and accessible mobility, develop green jobs and greater local self-reliance, nurture a local circular economy, and minimize environmental impacts. However, it requires rethinking mobility from the ground up in terms of products, manufacturing processes, business models, and supportive civic infrastructure. The effectiveness of these policies will vary depending on regional factors and the overall economic and political climate.

9.6.
An Opportunity for Automakers

Established automakers may be forced to rethink their approach to serving people who live in cities, which are becoming more congested and environmentally conscious, and which need a more seamless integration between personal vehicles and public transportation. This could involve developing cars that can easily integrate with public transport systems, creating multimodal transportation solutions, and investing in carsharing and ride-sharing technologies. At the same time, oversized vehicles, particularly in North America, may not be financially sustainable in the long term, and automakers may need to develop more efficient, compact vehicles suitable for urban environments and to offer a wider range of vehicle sizes to meet diverse consumer needs, emphasizing the benefits of smaller vehicles in terms of cost, environmental impact, and urban practicality.

The development and sale of micro-EVs could actually be good business for existing automakers who do not typically sell many vehicles to people living in dense, urban areas anyway because vehicle ownership tends to decline with higher population density, as discussed in Chapter 2, Section 2.3.1. Automakers already have the engineering expertise and need only apply it to a smaller form factor. Therefore, the premise behind this being an opportunity is that it can leverage existing competencies to generate incremental sales, rather than cannibalizing existing sales. Rivian, for example, is embracing this prospect and has spun off Also, Inc. as a startup to focus on micromobility [9.13].

This is even more the case if urban ride-hailing fleet vehicles and goods delivery vehicles are included as part of the opportunity. There are an estimated 95 ride-hailing vehicles and 22 taxis for every 10,000 inhabitants across a wide variety of cities worldwide [9.14], which implies that there are roughly 30 million taxis and ride-hailing vehicles in the world. There could be a similar number of urban delivery vans worldwide, and if these 60 million vehicles are replaced every four years on average because of their high mileage, it is possible that the annual total addressable market (TAM) for 20 million urban vehicles that move both people and goods could be around $600B, assuming the average cost of the vehicles is $40,000. Statistics are difficult to find and can be dated, but a 2016 Pew Research Center study found that 21% of city center residents had used ride-hailing services, versus 15% of suburbanites, and just 3% of rural residents [9.15]. Assuming that around 25% of the TAM applies to vehicles that operate primarily in the city center, such as micro-EVs and robopods, then the annual opportunity for city-center microvehicles could reach $150B. An automaker grabbing a 10% share of this "pie" could increase its revenues by more than 10% and with good potential for repeat business from fleet customers. Making the market even more attractive is the forecast that last-mile delivery is expected to grow 78% between 2023 and 2030, which could lead to a projected 36% increase in the number of delivery vehicles [9.16]. With trends in urbanization (Chapter 2, Section 2.4.1) and affordability (Chapter 5, Section 5.7), it is reasonable to assume that the market for vehicles to support ride-hailing, last-mile delivery and dense, urban mobility will grow at a significantly faster rate than the general automotive market.

To recap from Chapter 2, Section 2.3.1 there are around 500 cities in the world with a population of at least one million. Using the existing density of taxis, ride-hailing, and urban delivery vehicles, there could be an opportunity to replace around 5000 city center vehicles every four years for each city. A ten-year supply contract could reach $500M for such a city and exceed $10B for some megacities for vehicle purchases alone with the opportunity for ongoing OTA update revenues. A collaboration between automakers and local authorities could lead not only to tailored vehicle designs that meet specific local or regional needs, but also to the development of infrastructure to support smart EVs (e.g., charging stations, V2I communications), and to policies that encourage the adoption of more sustainable transportation options.

These changes would represent a significant shift in how established automakers operate, moving from a globalized, "one-size-fits-all" approach to a more localized, adaptive strategy, and this will create new challenges. There is a risk, after all, that locally tailored vehicles might not achieve sufficient sales volume to justify the cost of developing different vehicle variants for different markets. A substantial investment may also be required to research, develop, and promote vehicles tailored to various local needs, and using local materials could reduce economies of scale. Shifting to local sourcing would require rebuilding supply chains in each market, further increasing cost and complexity. Organizationally, the company's culture would need to evolve from a centralized, global mindset to one that empowers local teams and diversity, with potential challenges for finding and retaining talent with local expertise in various markets, while maintaining a cohesive global workforce with effective knowledge sharing. This balance between being flexible enough to accommodate local variations while maintaining global standards also impacts areas such as manufacturing and quality control. A compromise might be for a microfactory in one location to export additional micro-EVs to other regional locations because most cities face the same fundamental traffic challenges and are looking to make their public transport system more accessible and cost-effective. This would generate greater economies of scale but with some loss in local customization.

This transition would be complex and potentially risky for any automaker, but it could also position an automaker to be more resilient and relevant in an increasingly diverse and localized global market. The key would be to find the right balance between global efficiency and local adaptability. If an automaker is reluctant to reinvent its business model and set up microfactories, there is a less radical option, which is to mass-produce a standardized ACE platform and to supply it, or license its design, to microfactory operators around the world. This would provide economies of scale for the most challenging and expensive part of the vehicle, comprising the propulsion system, chassis frame and systems as well as the autonomy compute platform; there is, after all, significant technology dual use with conventional automobiles on vehicle autonomy, electrification, connectivity, and software. The microfactory operator could then focus on designing and producing the coach and user experience, using locally available and renewable materials. In effect, the automaker would supply the platform and would choose whether to collaborate or invest on the coach side with funds or know-how, such as software and infotainment, as well as supply relationships. The automaker may also choose to be creative with supplying the ACE platform at cost but with access to the data generated by the vehicle fleet and with the opportunity to supply OTA updates throughout the vehicle's lifecycle.

Compared to traditional, mass-produced vehicles, this ACE platform with a microfactory coach build has lower capital investment, faster product iteration cycles, reduced inventory costs, and more predictable municipal procurement. It could allow automakers to achieve a first-mover advantage in urban mobility, enhance their brand reputation for innovation, and provide potential global scalability and a hedge against declining personal vehicle sales.

References

9.1. Battery and Electrification Technology, "Where Do EV Batteries Go When They Die," November/December 2024, https://www.techbriefs.com/component/content/article/52138-where-do-ev-batteries-go-when-they-die.

9.2. Mercedes-Benz Media UK, "Pioneering Innovations for the Car of the Future: Mercedes-Benz Provides Exclusive Insights into Research Activities and Future Technologies," Mercedes-Benz Press Release, November 22, 2024, https://media.mbusa.com/releases/

pioneering-innovations-for-the-car-of-the-future-mercedes-benz-provides-exclusive-insights-into-research-activities-and-future-technologies.

9.3. Wikipedia, "Zabbaleen," https://en.wikipedia.org/wiki/Zabbaleen.

9.4. Stumpf, R., "BYD Hit $100 Billion. Next: An EV Factory Bigger than San Francisco," InsideEVs, March 24, 2025, https://insideevs.com/news/754460/byd-100-billion-huge-factory/.

9.5. M-Kopa Website, "We Finance Progress," https://www.m-kopa.com/about.

9.6. Transport for London, "Transport for London Quarterly Performance Report: Quarter 1 2023/24 (1 April – 24 June 2023)," https://content.tfl.gov.uk/tfl-quarterly-performance-report-q1-2023-24-acc.pdf.

9.7. Shepardson, D., "New York to Launch $9 Manhattan Congestion Charge in January," Reuters, November 14, 2024, https://www.reuters.com/world/us/new-york-moving-revive-manhattan-congestion-charge-2024-11-14/.

9.8. Goldstein, E., "Busting the Myths of New York's Congestion Pricing Program," NRDC, December 26, 2024, https://www.nrdc.org/bio/eric-goldstein/busting-myths-new-yorks-congestion-pricing-program#:~:text=Myth%20%232%3A%20Congestion%20pricing%20is,from%20the%20Central%20Business%20District.

9.9. Greater London Authority, "London-Wide Ultra Low Emission Zone – One Year Report," March 2025, https://content.tfl.gov.uk/london-wide-ulez-one-year-report.pdf.

9.10. Global Traveler website, "Helsinki to Go Car Free By 2050," January 13, 2016, https://www.globaltravelerusa.com/helsinki-to-go-car-free-by-2050/.

9.11. New York City MTA Website, "Subway and Bus Ridership for 2023," https://new.mta.info/agency/new-york-city-transit/subway-bus-ridership-2023.

9.12. Metropolitan Transit Authority, "November 2024 Financial Plan Presentation," November 20, 2024, https://new.mta.info/transparency/financial-information/financial-and-budget-statements.

9.13. Rivian Website, "Rivian Spins out Micromobility Business into New Startup—Also, Inc." March 26, 2025, https://rivian.com/newsroom/article/rivian-spins-out-micromobility-business-into-new-startup-also-inc.

9.14. UITP Advancing Public Transport, "Global Taxi & Ride-Hailing Figures 2024," September 2024, https://www.uitp.org/publications/global-taxi-ride-hailing-figures-2024/.

9.15. Smith, A., "On-Demand: Ride-Hailing Apps," Pew Research Center, May 19, 2016, https://www.pewresearch.org/internet/2016/05/19/on-demand-ride-hailing-apps/.

9.16. World Economic Forum, "Zero-Emission Delivery Is Possible for Retailers – Here's What's Needed," September 18, 2023, https://www.weforum.org/stories/2023/09/zero-emission-last-mile-delivery-ikea-e-cargo-bike/.

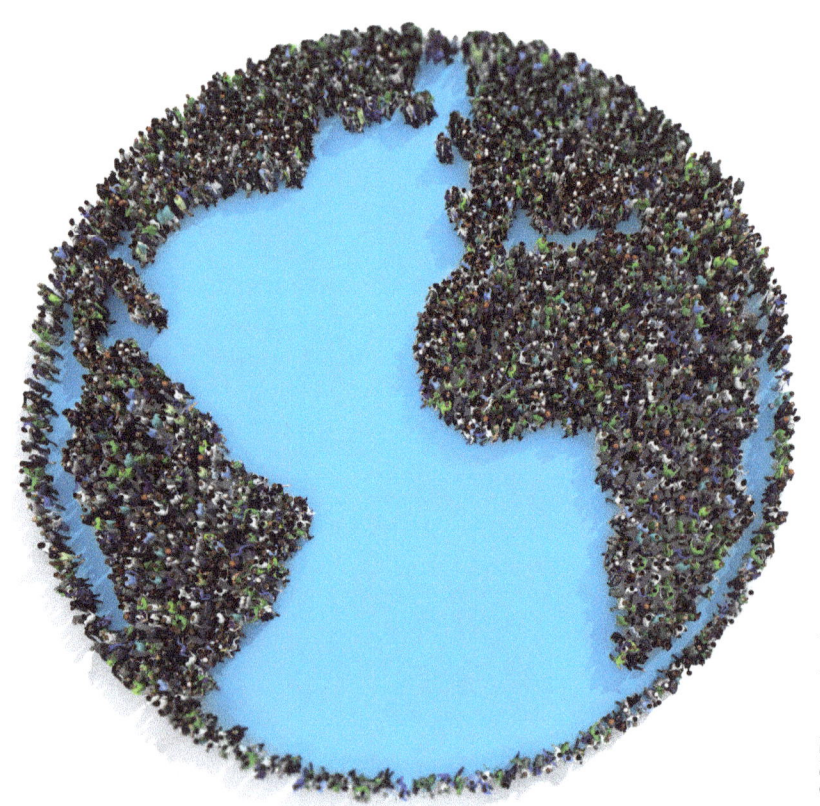

Chapter 10

Conclusion

10.1. Rethinking the Future of the Auto Industry

The auto industry is going through a revolution, not of its own making, but forced on it by three Silicon Valley companies around 2010—Tesla, Uber, and Alphabet (Waymo). These companies made the early running in EVs, ride-hailing, and AVs, but their strongest competition now comes from China. Technology companies, such as Amazon, Nvidia, Apple, Huawei and Baidu, are also investing heavily in AVs, in particular, for two main reasons. The first is that it is a very difficult software problem to solve, requiring deep pockets, but if AVs can be commercialized, then the underlying solutions can probably be applied to many other industries. The second is that there is likely to be a lot of money that can be made from eliminating the need for driving (in personal vehicles) and the need for drivers (in commercial fleet vehicles).

But who is thinking about what people and society really need? Connected vehicles, for example, are attractive to tech companies and automakers because they allow them to gather more data on vehicle and customer usage and sell more data, services, and OTA updates, but the downsides are significant for the customer in terms of the potential loss of privacy, security, and even vehicle functionality or access in the long term. Meanwhile, EVs should be friendlier to the planet if they use recycled batteries and if the electricity generated is increasingly green, but if they weigh more than ICEVs, they can pose a greater risk to other road users, decreasing overall road safety and increasing vehicle-related fatalities even if they are equipped with ADAS. It can also increase wear and tear on the roads, bridges, and parking lots. There seems to be a lack of "joined up" thinking because vehicles are treated by regulatory agencies as standalone objects even though their environmental impact goes far beyond vehicle operation into the cloud (servers) and the earth (mines), and their safety performance is viewed only from the perspective of the vehicle occupants and not in terms of overall road safety.

There is an undeniable "dark" side to many of the technologies being developed in the auto industry due to concerns for privacy, security, job loss, human rights, environmental degradation, and so on. To some extent, this is because the industry sees mobility problems as a compelling technical or software challenge to solve instead of solving the real needs or problems, which are more systemic and require greater collaboration between stakeholders. As an example, if drivers are distracted, then the prevailing auto industry logic, endorsed by national or regional governments, is that driver-monitoring technology should be added when the logical solution would be to reduce the source of distraction in the first place. Unfortunately, making it impossible for the driver to use a smartphone in the vehicle is not considered an attractive solution when so much effort has been made to enable the driver to take their hands off the wheel and feet off the pedals just to stare at the road ahead!

10.2.
A Vision for Integrated Mobility

Fundamentally, people need affordable and convenient solutions to move around, and society needs mobility to be sustainable and to serve everyone, but lip service is being given to making society a better place.

In theory, road safety could be improved, mobility would be easier for the disabled, and vehicle energy usage might be reduced with AVs. However, there are many reasons to believe why this will not happen because, compared to a professionally driven ride-hailing vehicle, a robotaxi may offer no overall improvement in road safety, may be less able to help a disabled person enter and exit the vehicle, and may end up consuming more energy because of its compute needs in the vehicle and in the cloud.

Fortunately, the very same technologies that are being developed by tech companies and automakers can be deployed to support cities in transforming urban mobility. Autonomous operation, for example, assisted by connectivity (to help with routing, scheduling, pickup, and so on) and electrification (to power the autonomy software stack and to support city air quality objectives), can be used to provide last-mile transport, addressing the weak points of public transport and providing people with the "Holy Grail" of affordable, on-demand, door-to-door mobility, without needing to find a parking space! The vision is that by complementing or, in some cases, replacing individual car ownership with a comprehensive MaaS, we may arrive at a more convenient, sustainable, and affordable way to "go wherever I want, whenever I want" and offer it to a much broader group of people. The key is to view transportation not as a set of independent solutions, which is the case now, but as an integrated system that leverages the different strengths of bicycles, buses, and trains and combines them with right-sized vehicles so that it can serve everyone and can dynamically adapt to changing needs far easier than building new roads or lanes.

10.3.
An Opportunity for the Auto Industry

This vision need not be a threat to the auto industry. On the contrary, it can be a much-needed opportunity to augment their existing business. After all, the auto industry is facing the three technology challenges of autonomy, connectivity, and electrification at the same time as its business model is being attacked on three fronts as well. First, their automobile product is ill-suited to cities where more of the world's population is moving to and living, and where there are challenges with congestion, parking, road safety, air pollution, and accessibility. A truly "smart" city should focus on quality of life, which may not be the same as making life easier for car users! Second, vehicles are becoming increasingly unaffordable to purchase, insure, and repair for the average person. Third, they are becoming uncompetitive with those developed by Chinese companies, who are not only offering far more affordable EVs but are also leading the world in battery, charging and swapping innovation.

These three developments put at risk a significant driver of the developed world economies and a significant source of employment for an industry that already needs to adapt to new technologies like simpler electric powertrains, AI, and autonomous operation, each of which can lead to fewer good-paying jobs. But some cities are looking to restrict cars or their speed, and this can lay the groundwork for a new type of low-speed EV that may not have to meet all the safety standards, could be made from a wider choice of materials, and could be made locally and in every city around the world! As a response, traditional automakers may need to give more consideration to right-sizing vehicles, offering subscriptions and flexible ownership models, fostering a closer integration with cities and their mobility services, and partnering with local microfactories. The most successful automakers will likely be those that can create affordable ACE platforms or skateboards that can be quickly modified for different urban markets worldwide to provide effective solutions for the last-mile movement of people and goods.

10.4.
A Sustainable Future for All

Encouragingly, the specific enabling technologies—batteries, solar panels, cameras, semiconductor chips, and wireless connectivity—are rapidly falling down the cost curve and becoming widely available all over the world. This means that advanced mobility solutions can become feasible even for the poorest regions of the world, which can also include parts of some cities and rural areas that are in wealthy countries.

This may take the form of shared mobility solutions and E2W/E3W or just an e-kit that can be retrofitted to commonly used nonmotorized vehicles like tricycles, handcarts, and wheelchairs. Such a solution may also be useful for applications in the "rich" world, such as last-mile goods delivery, indoor healthcare and hospitality facilities, military bases, farms, and so on.

In the opposite direction, approaches that are more commonly seen in poorer regions are destined to become more practical in wealthier countries. An example is increased recycling and use of natural materials, like wood, bamboo, and hemp, because the strict materials performance requirements for automobile crashworthiness and comfort can be relaxed for the type of low-speed microvehicles needed for "last-mile" urban operation. Right-sized vehicles made of recycled and natural materials and having a solar panel roof can be used to complement public transport and make moving people and goods easier on the wallet and on the planet. Producing these vehicles in local microfactories can also create good jobs to offset those lost through automation, tackle urban solid landfill waste, increase civic pride and engagement, and help make communities more self-sufficient.

Instead of the angst surrounding cybersecurity hacks, AI dystopia, and robots "run amok," promoted and developed by powerful tech companies, what is being proposed here is to put the "heart" back into technology by focusing more on the needs of people and society. Putting the end user at the center, not the technology, and providing sustainable and affordable transport solutions for everyone—what a novel concept!

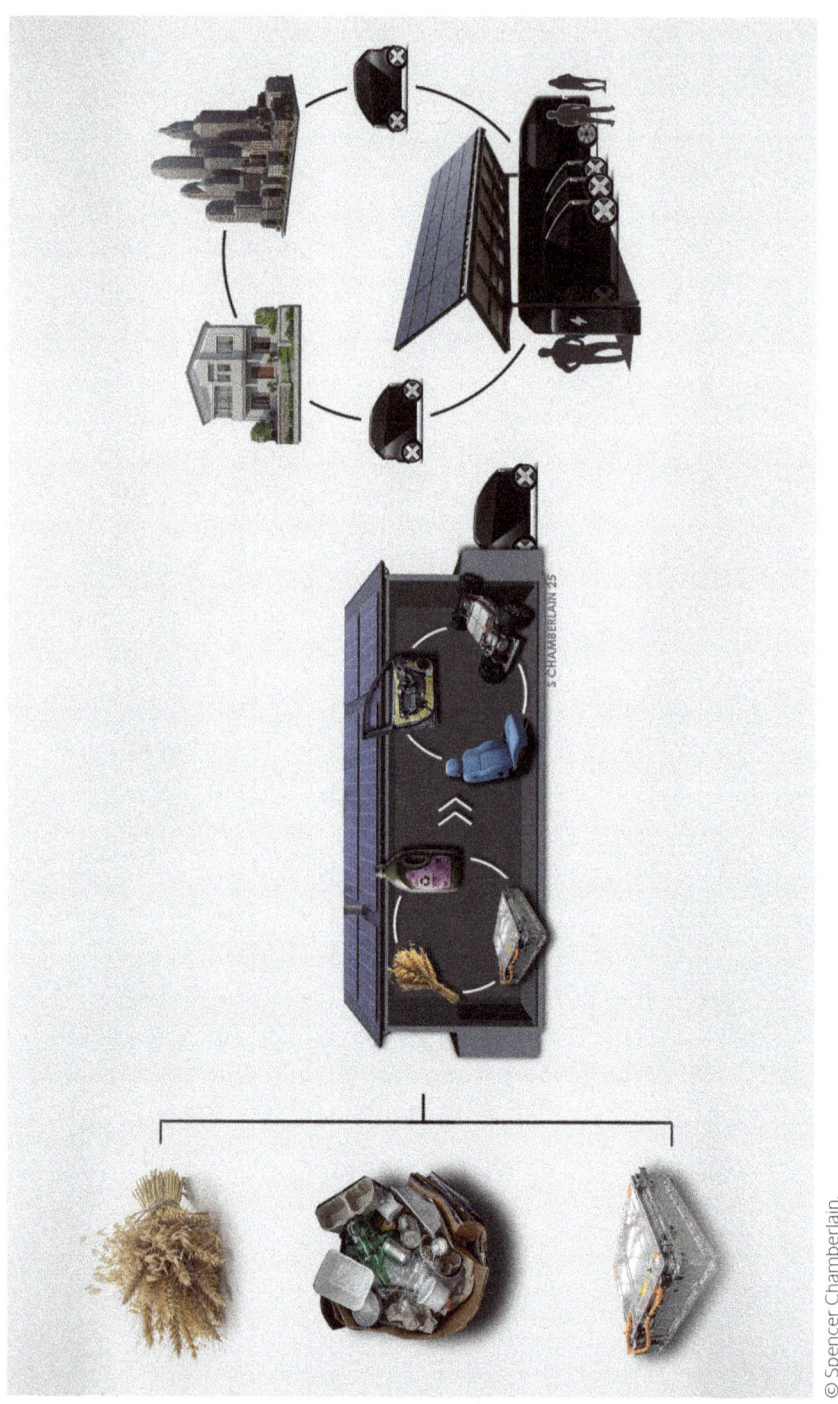

As discussed in the book, locally available materials and recycled parts can be the feedstock from which "right-sized" vehicles can be locally made, which then provide last-mile connectivity with public transport. Creating jobs, tackling waste and improving access and mobility for the underserved equals *Sustainable and Affordable Mobility for All: Putting the Heart back into Technology.*

© Spencer Chamberlain.

Afterword I

The author of this book is a visionary mind, offering us a profound opportunity to reflect on the challenges of modern mobility and the path we are taking to solve them. Today, it is easier to think of solutions through incremental innovation—layering more and more technology over existing frameworks without addressing the root causes of our mobility issues. But true transformation requires a different approach; mobility is a complex system, and despite a century of advancements, its fundamental structure remains largely unchanged. What if, instead of simply optimizing the current system, we reimagined it from its foundations, strategically applying innovation where it matters most?

We stand at a crossroads in the mobility industry, where technology, public policy, entrepreneurs, and disruptive thinkers from various industries have an opportunity to challenge the status quo. The choices we make today will not only determine how we move but also redefine our relationship with the environment, urban spaces, and each other. The need for transformation is evident, but so is the opportunity to create vehicles that are more than just means of transport—they can be catalysts for a sustainable, efficient, and equitable future.

A key part of this shift is the adoption of light EVs. I firmly believe in their necessity for several reasons: they are vastly more energy-efficient, allowing us to rethink vehicle architecture and materials in ways that harmonize with nature rather than exploit it. They present an economically viable option for a significant portion of the global population, who might otherwise have no access to personal mobility. Additionally, they address urban challenges such as traffic congestion and parking constraints. More than that, they offer the chance to redefine the consumer's perception of vehicles—moving beyond aesthetics and status symbols to embrace mobility as a tool for progress, a work asset, and a business enabler.

This book has taken us on a journey through groundbreaking concepts that challenge conventional automotive thinking. It compels us to ask fundamental questions: How can we reduce energy consumption tenfold? How can vehicles integrate with nature's principles of efficiency and adaptability? How do we ensure mobility solutions cater to diverse global needs, from dense urban centers to rural communities?

At the heart of this revolution is a shift in mindset—from designing vehicles as standalone entities to seeing them as interconnected components of a larger system. The traditional approach has long prioritized more power, faster acceleration, and increased complexity, often at the cost of sustainability and accessibility. But the future demands a new paradigm, one that considers the broader implications of vehicle use, resource consumption, and environmental impact.

One of the most exciting areas explored in this book is the role of biomimicry in vehicle design. Nature's solutions—refined over millions of years—offer incredible insights into energy efficiency, durability, and adaptability. From the aerodynamic efficiency of marine life to the self-repairing properties of bone, we can draw inspiration to create vehicles that are lighter, stronger, and more in sync with the world around them. Modularity, energy harvesting, and connected platforms further enhance the potential for truly transformative mobility solutions.

Beyond the technical and ecological innovations, this book challenges us to rethink our societal values around mobility. It emphasizes the importance of right-sizing vehicles to match real-world needs, reducing excess, and prioritizing purpose over prestige. It advocates for solutions that tackle urban congestion, energy scarcity, and climate change while ensuring mobility remains inclusive and accessible to all.

For me, this is more than just theory—it is a mission. That is why, in 2024, I embarked on my own moonshot project: developing the first Mexican light EV and its future mobility ecosystem. This is my contribution to a future where mobility is redefined, where efficiency and accessibility go hand in hand, and where innovation is strategically applied to create real impact.

This afterword is an invitation to think boldly, act courageously, and dream expansively. The future of mobility is not a distant vision—it is being built today, shaped by the decisions we make and the systems we create. As you delve into the pages of this book, I encourage you to embrace its vision and contribute your own ingenuity to this transformative journey. Together, we can design a future where vehicles are not just tools but partners in building a sustainable, connected, and inspiring world.

Ricardo Daniel Apaez Pérez
CEO, DRIVEN/CLAUT Innovation Center
Mobility Tech Advocate & Entrepreneur
Empowering humanity by making mobility an enabler of progress

Afterword II

I am the Founder CEO of EVage Motors, an Indian company that has spent the last decade developing purpose-built commercial EV's fusing automobile design with aerospace engineering to reimagine and transform the mobility industry. We have developed breakthrough products using a modular structure that can support many vehicle types, with a lightweight and strong exoskeleton covered with space grade materials. Our vehicles use ultra-safe and fast charging batteries and are manufactured in our highly efficient Modular Micro-Manufacturing factory.

Chris Borroni-Bird needs no introduction in the automotive world as he is gifted with a technical brain that is programmed to think differently, evident from his career and the vehicle platforms and technologies he developed while working with well-known automotive and technology companies. Beyond the obvious innovations, one can find all the developments undertaken by him are deeply rooted in making the world more sustainable and a better place for the generations ahead. Chris' work on the Afreecar project also highlights the impact technology can have in society to offer a better livelihood for people who are in need of tech the most.

As always the book aims at creating a vision for the future of the automobile, aptly pointing out what is wrong with the industry and what needs to be done to refresh the modern automotive industry. This is sorely needed at this juncture because the major automakers are struggling to find a footing with increased competition from China and the changing needs of customers and regulators around the world.

What I appreciate most about the book is its holistic focus on improving the vehicle through both software and hardware and the development of the ACE platform, which could be a way forward to create synergies between tech companies and the automotive industry as a whole. A deeper understanding of the use of lightweight bio-materials and nature-inspired aerodynamic solutions are concepts that should be visible in future automobiles. The biggest shift predicted and the most critical point, in my opinion, is the right-sizing of vehicle platforms which is still hidden away under the garb of creating more powerful vehicles. As pointed out in the book, the need to create innovative manufacturing solutions will also be a huge tectonic shift aided by the new-age materials, nature-inspired design, and right-sizing.

This book raises pertinent questions that plague the automotive industry but it also lays down the path for the next few decades by creating a playbook for designing and manufacturing future automobiles. I highly recommend this book to academics, urban planners, policy makers, designers, engineers, and C-suite executives looking to reinvent the mobility landscape of the world.

Having worked with Chris over the last five years, there is a deep sense of respect and admiration for him, and a lot of what we do at EVage Motors is inspired by Chris's work as an automobile guru. Behind the technical genius Chris is a great human being with strong values.

I hope this book Inspires the next generation of automotive change-makers!

Inderveer Singh Panesar
Founder and CEO, EVage Ventures

Index

3D printer, 83
3D printing, 308–311

A

Abundant storage space, 10
Active mobility, 201
Adaptive cruise control (ACC), 19
Advanced air mobility (AAM), 130
Advanced driver assistance system (ADAS), 19, 154–155, 330
  advanced, 155
  basic, 155
  cost, 155–157
  effectiveness, 157–160
  features, 158
  mid-level, 155
Advanced insulation, 249
Advanced restraint systems, 18
Aerodynamic improvements, 248
African innovation, 83–85
Agbogbloshie, 37
Age demographics, 29–30
Airbag control modules, 18
Air resistance, 248
Air taxis, 128–132
All-wheel drive (AWD) vehicle, 103
Alternating current (AC), 45
American Automobile Association (AAA), 156, 158
American Automobile Association (AAA) Foundation for Traffic Safety, 8
American automotive innovation, 180
Americans with Disability Act (ADA), 163

Amplitude modulation (AM) radio, 17
Antilock braking system (ABS), 18
Artificial intelligence (AI), 121
Artificial intelligence (AI)-powered voice assistants, 119
Audi integrated lidar, 109
Augmented reality, 119
Australian government, 189
Auto industry, 329–332
Auto insurance rates, 172
Automated charging, 94
Automated guided vehicles (AGVs), 281
Automatic emergency braking (AEB), 19
Auto microfactory, 306
Auto-mobility, 3, 179, 205
  automotive technology, 180–181
  autonomy and public transport, 220–227
  emerging countries, in EV value chain, 189–194
  goods transport, 236–241
  hierarchy of new framework, 200–207
  "inevitable" EV transition, 181–186
  MaaS, 227–231
  major automakers, rise of EVs, 186–188
  potential role for city government, 242–243
  removing "friction" in vehicle subscription service, 231–236
  traditional automakers, 195–200
  urban design for mobility, 209–220

Automotive electronics, 16–19
  history, 17
  life of typical car, 19–22
  motorized passenger vehicles, 23–25
Automotive (r)evolution
  autonomous, connected, electric mobility, 73–75
  autonomous vehicle, 66–73
  connectivity and software-defined vehicles, 62–66
  electric vehicle
    BYD Seagull interior, 54
    Chevrolet Volt, 52, 53
    evolution of battery technologies, 44
    GM's Autonomy, 45, 48
    GM's EV1 two-seater, 44–46
    GM's Hy-wire, 45, 49
    ground-up, 47
    modern EVs, 45
    Nissan Leaf, 50
    technology, 57–62
    Tesla Model 3, 54
    Tesla Model S, 52, 54
    Tesla Roadster, 49
    Toyota Prius, 45, 47
  microvehicles, 55–57
  mobility-as-a-service, 75–82
Automotive manufacturing, 297
  plant complex, 294
  processes, 309
Automotive material, 302
Automotive plastics, 299
Automotive recycling, 298
Automotive research, 249

© 2025 SAE International

Automotive trajectory
  ADAS, 154–155
    cost, 155–157
    effectiveness, 157–160
  AVs
    accessibility, 162–164
    congestion, 164–165
    energy usage, 166–168
    safety, 160–162
    workforce and economy, 169
  connected and SDVs, 147–148
    life cycle concerns, 152–153
    privacy, 148–150
    security, 151–152
  EVs
    ESG issues, 141–142
    mass issues, 145–147
    national security issues, 142–145
    skeptics, 140
  systems thinking, 139–140
  unintended consequences, 135–139
  vehicle affordability, 170–173
Autonomous delivery vehicles, 238
Autonomous Mercedes-Benz S-Class, 68
Autonomous operation, 331
Autonomous shuttle, 223
Autonomous truck sensor suite, 237
Autonomous vehicle (AV), 5, 255, 329
  accessibility, 162–164
  appropriate maneuvers, 67
  business goal, 114
  cameras, 70
  Chinese government, 118
  city traffic management operations, 114
  compute platform, 110, 111
  congestion, 164–165
  data collection, 67
  energy usage, 166–168
  estimates for robotaxi economics, 117
  insurance costs, 116
  levels of driving automation, 108, 109
  lidar, 71
  low-latency communications, 118
  low-level control, 67
  machine learning progress, 72
  modular, end-to-end and hybrid AV software approaches, 112
  Project Prometheus, 67
  radar, 70–71
  safety, 160–162
  software development, 118
  space, 110
  technical challenges, 115
  technical reason, 111
  typical sensors and sensor locations for AV operation, 71
  vision, 66
  workforce and economy, 169
Autonomy, 78
Autonomy, connectivity, and electrification (ACE), 307, 313, 326
  e-kit, 279–283
  reference design, 273–278
Auto parts market in Accra, Ghana, 36
AutoPilot, 18
Avis, 76

B

Balanced approach, 206
Bamboo, 303
Battery charging
  automated charging, 94
  bidirectional charging capability, 96
  daytime charging, 95
  electric grid and renewable energy storage, 95
  ultra-fast charging, 92, 93
  wireless charging, 93, 94
Battery EVs (BEVs), 5
Battery management system (BMS), 57
Battery materials, 58
Battery performance, 89–92
Battery recycling, 144, 145
Battery technology roadmap, 89, 90
Bio-based materials, 248, 302
Biometric facial recognition, 122
Blind spot warning, 158
Blockchain, 128, 234
Bogotá TransMilenio bus rapid transit, 216
Brain-inspired computing technologies, 255
Build your dreams (BYD), 305
Bus rapid transit (BRT), 216
BYD Seagull interior, 55

C

Cairo, 301
California Air Resources Board (CARB), 195
Car2go, 76
Car ownership, 24
Carsharing services, 76
CATL, 59–61, 90, 268
Cell-to-pack (CTP), 144
Cellular communications technologies, 136
ChatGPT-4, 167
Chevrolet "Boss" Tahoe, 69, 70
Chevrolet Volt, 52, 53
Chilean government, 189
China Light-Duty Vehicle Test Cycle, 60
Chinese automakers, 119, 181
Chinese Test Cycle, 54
Circular economy, 297–301
Circular mobility ecosystem, 319
Citroen Ami, 264
Civil Aviation Administration of China (CAAC), 131
Clean Air Acts, 18
Cloud-based storage, 151
C-Max Solar Energy Concept, 259
CO_2 emissions, 12
Commercial transport, 236
Commercial vehicle, 240
Conductive back sheet, 257
Congestion charge, 314, 315

Sustainable and Affordable Mobility for All **343**

Connected vehicles, 62–64, 329
Conventional auto plants, 307
Conversion efficiency, 250
Convolutional neural networks (CNNs), 72
Corporate Average Fuel Economy (CAFE) standards, 11
COVID-19, 171, 198
Cruise Control, 18
Cyber-bullying, 136
Cybersecurity, 19, 64, 128

D

Daimler Victoria, 75
DARPA Grand Challenges, 69
DARPA's Urban Challenge, 69, 70, 196
Democratic Republic of the Congo (DRC), 191
Deterministic AI, 111
DiDi, 77
Direct current (DC), 60
Distributed ledger, 128
Door-to-door transport service, 224
Driver errors, 154
Driver feedback systems, 252
Driver monitoring system (DMS), 121, 122, 158, 159
Driver surveillance, 159
Driver wages, 237
Dynamic routing software, 238, 240

E

Early Tesla models, 45
E-cart, 288
E-commerce, 212
Economic policy, 28
E/E content, 18, 19
E-kit, 279–283, 289, 290
Electrical gauges, 17
Electrical systems, 249
Electric three-wheelers (E3W), 55
Electric two- and three-wheelers (E2W/E3W), 263, 264
Electric two-wheelers (E2W), 55
Electric vehicle (EV), 332

annual global automotive sales and shift, 182
automakers, 95
battery supply chain, 192
bidirectional charging capability, 96
BYD Seagull interior, 55
charging infrastructure, 183
charging port, 94
Chevrolet Volt, 52, 53
China's emergence, 183
Chinese EV startups, 184
Chinese government policies, 184
cumulative investment commitments, 188
domestic and international markets, 185
doubling EV range on cost of ownership, 266
emerging countries, in EV value chain, 189–194
ESG issues, 141–142
evolution of battery technologies, 44
GM's Autonomy, 45, 48
GM's EV1 two-seater, 44–46
GM's Hy-wire, 45, 49
ground-up, 47
"inevitable" EV transition, 181–186
lineups, 186
major automakers, rise of EVs, 186–188
major sources for lithium-ion battery EV materials, 190
mass issues, 145–147
mass-market, 90
micro-EVs, 204
modern EVs, 45
national security issues, 142–145
new electric mobility ecosystem, 97
Nissan Leaf, 50
personal, 115
power electrical appliances and tools, 95
projected EV sales, 280
right-sized, 263
skeptics, 140

technology, 57–62
Tesla Model 3, 54
Tesla Model S, 52, 54
Tesla Roadster, 49
Toyota Prius, 45, 47
typical specifications and current cost estimates, 56
wireless EV charging, 93
Electric vertical takeoff and landing (eVTOL), 128–132
Electrification, 220, 226, 331
Electronic control units (ECUs), 18
Electronics content, 107
Electronic stability control (ESC), 18
EMBARQ India's analysis, 24
Emerging automotive landscape
AVs, 108–119
battery charging, 92–99
battery performance, 89–92
cabin experience, 119–124
chassis and propulsion, 101–106
eVTOLs (air taxis), 128–132
hydrogen fuel cells, 99–101
new automotive E/E architecture, 106–108
V2X, 124–128
Employment, 318
End-to-end AI, 113
Energy consumption, 248
Energy content, 58, 59
Energy conversion and storage, 254
Energy efficiency, 74
Energy harvesting, 251
Energy source, 10
Energy storage systems, 253
Energy usage, 166–168
Environmental Protection Agency (EPA), 195
Environmental, social, and governance (ESG) issues, 141–142
E-trailer, 288
Euro New Car Assessment Program, 121
European Union Aviation Safety Agency (EASA), 131

E-waste site, 37
Extensive recycling, 301

F

Federal Aviation Authority (FAA), 130
Federal Highway Administration, 8
Federal Motor Carrier Safety Administration, 161
Federal Trade Commission (FTC), 148
2000 Ferrari Modena 360 sports car, 15
Financial Crisis, 180
Financial sustainability, 81
First-generation VW Golf, 13
Flax, 303
1974 Ford F-150, 13
Ford, Henry, 1, 198, 199, 293, 295
Ford Ranger, 13
Fuel cell electric vehicle (FCEV), 5, 99, 100, 270
Fuel economy, 11, 12, 15, 17, 138
Fulbright Scholarship, 33
Full self-driving (FSD), 107, 167
Fully automated vehicle, 5

G

Gasoline generator, 52
Gas Station, in rural Mali, 32
Glencore, 144
Global auto-mobility
  automotive electronics, rise of, 16–19
    life of typical car, 19–22
    motorized passenger vehicles, 23–25
  societal and demographic trends, 25
    age demographics, 29–30
    obesity, 31
    urbanization, 26–28
  vehicle usage, 7–9
    automobiles, overengineering, 11–16
    Maslow's hierarchy of needs, 9–11
Global economic dominance, 180
Global efficiency, 326
Global life expectancy, 30
Global lithium-ion battery supply chain, 142
Global obesity rates, 31
Glyde, 80
GM EN-V (Electric Networked-Vehicle) concept, 73, 74
GM Hy-wire, 45, 49
GM's Autonomy, 45, 48
GM's EV1 two-seater, 44–46
GM's Hydrotec fuel cell, 98
Goods transport, 236–241
Greenhouse Gases, Regulated Emissions and Energy Use in Technologies (GREET), 140

H

Hague's car-free city center, 317
Health monitoring, 150
Heating, ventilation and air conditioning (HVAC), 249
Hemp fiber, 303
Hertz, 76
High-power DC electricity, 98
Holographic windshield display (HWD), 121
Horse-drawn carriages, 137
Household income, 27
Human brain activity data, 256
Human error, 154
Human–machine interfaces (HMI), 19
Hybrid electric vehicle (HEV), 5, 45, 267
Hydraulic braking, 45
Hydrogen fuel cells, 99–101
Hyundai, 307
Hyundai Mobis, 121
Hyundai Motor Company, 256

I

Import tariffs, 171
Indonesia, 190
  EV-related foreign investment, 191
  nickel mining, 191
  workforce, 191
Inflation, 172
Infotainment systems, 18
Insurance Information Institute (III), 156
Insurance Institute for Highway Safety (IIHS), 157
Integrated mobility, 330–331
Internal combustion engine (ICE), 249
Internal combustion engine vehicle (ICEV), 16, 17, 43, 44, 50, 55, 58, 60, 62, 73, 140, 146, 181, 182, 251, 267, 298, 315, 330
Internal digital assistant, 121
Internal Revenue Service (IRS), 171
International Council on Clean Transportation (ICCT), 266
International Energy Agency (IEA), 11
International partner ships, 143
Internet of Things (IoT)-enabled smart lockers, 238
In-vehicle units (IU), 219
In-wheel motor, 103, 105, 106

J

Joint ventures (JVs), 183

K

Kei cars, 264, 265
"The Knowledge," 77

L

Lane departure warning (LDW), 19
Laptop batteries, 85
Large language model (LLM), 167
Leapmotor cars, 185
LED lighting, 284
Lidar, 71
Lightyear, 257
Lithium-based battery, 59
Lithium-ion batteries, 182
Lithium-ion battery EV materials, 190
Local data servers, 126
Local economic mobility system, 321
Localized production, 305–313
Logistics sector, 236
London Heathrow Airport's Ultra Pod PRT system, 225

Low-income communities, 212
Low-speed micro-EVs, 273
Lucid Motors, 194
Lyft, 77, 78, 115

M

Machine learning, 92, 111
Mali
   Gas Station, 32
   volunteering experiences in, 284–289
Manufacturing job, 318
Market segmentation, 1
Maslow, Abraham, 9
Maslow's hierarchy of needs, 9–11
Mass production, 295
Mechanical fuel carburetor, 18
Median household income, 172
Medical supplies, 241
Mercedes-Benz Vario van, 67–68
Mercedes S-Class, 67
Metals recovery, Agbogbloshie e-waste site in Accra, Ghana, 37
Metropolitan Transit Authority, 315
Metropolitan Transportation Authority (MTA), 319
Micro-EVs, 204, 272, 317, 320, 322, 323, 325
Microfactories, 307, 308, 311, 312
Micromobility, 30, 79–81, 90
Micron-thick photovoltaic films, 258
Microvehicles, 55–57
Mid-level vehicle, 20
M-Kopa, 282
Mobility-as-a-service (MaaS), 201–203, 227–231
   digital innovations, 81
   MaaS app, 81, 82
   micromobility, 79–81
   multimodal transport, 82
   and public transport integration, 78, 79
Mobility services, 79
Modern full-size pickup trucks, 13
Motivational theory, 9
Motor cost, 102
Motorized passenger vehicles
   car ownership, 24
   global motorcycle revenue, 24
   global sales of motorbikes, 24
   government registration data, 24
   motorbike usage, 25
   2024 Sales of motor vehicles, in India, 23
   tuk-tuks, 24
Motor vehicles, 137
Mozilla Foundation, 148

N

National Highway Traffic Safety Administration (NHTSA), 122
National security issues, 142–145
Natural gas, 249
Natural materials, 301–305
Nature-inspired automotive design, 253
Navigation systems, 251
Neodymium, 52
New automotive E/E architecture, 106–108
New York City Department of Transportation, 8
Nickel metal-hydride (NiMH) batteries, 45
Nio, 60, 61, 194
Nissan Leaf, 50, 96, 256
Nissan's PIVO 3, 74
Nonmotorized vehicles (NMVs), 279, 280, 282, 283, 289, 290

O

Obesity, 31
Occupant monitoring, 124, 150
Off-road military vehicle, 250
Oil Crisis, 18
On-board diagnostics (OBD), 18
OnStar, 63
Organization of the Petroleum Exporting Countries (OPEC), 142
Original equipment manufacturer (OEM), 199, 206
Over-the-air (OTA), 19, 64, 66, 73, 75
OX vehicle, 21, 22

P

Passenger vehicle, 240
Peer pressure, 233
Peer-to-peer sharing, 78
Permanent magnet (PM) motors, 52
Personalized preferences, 234
Personal mobility, 26
Personal rapid transit (PRT), 225
Pew Research Center, 324
Photovoltaic films, 258
Piezoelectric materials, 250
Piezoelectrics, 250
Plastic lead acid battery containers, 37
Platooning, 221, 222
Plug-in hybrid electric vehicle (PHEV), 5
Population density, 27, 28
Portugal tax new car registration, 13
Power electronics, 49
Powertrain efficiency, 251
Predictive maintenance, 233
Privacy, 148–150
Private transportation, 78
Projected EV sales, 280
Project Prometheus, 67, 68
Propulsion system, 10
Proton exchange membrane fuel cell (PEMFC), 99
Public engagement, 220
Public transport, 28, 30, 78, 79, 203, 206, 210, 212, 314
   and autonomy, 220–227
   infrastructure, 204
   integration, 79

Q

Qatar, 194
QuantumScape, 194

R

Radar, 70–71
Raw nickel ore, 190
Real-time tracking, 238
Regenerative braking, 250
Relative cost per weight, 170
Remote operators, 115
Remote working, 2
Renault, 223
Rental car services, 76
Residential streets, 146
Ride-hailing, 76, 77, 222
Road congestion, 239
Robopods, 224
Roboshuttles, 223, 224
Robotaxi business model, 116
Robotaxi economics, 117
Robotaxi operators, 104
Robotaxis, 131, 162, 164, 220, 224, 232, 235, 243
Robo-X, 224
Roof-mounted solar panels, 286

S

Samsung SDI, 60
Saudi Arabia, 193–194
Security, 151–152
Self-actualization, 10
Self-esteem needs, 10
Semi-autonomous systems, 159
Semi-complete knockdown, 21
Sensing technology, 69
Shared e-kit, 281
Shared mobility, 75, 333
SIXT, 77
Smart Driver program, 148
Smartphone E/E architecture, 107
Smartphone on wheels, 137
Smartphones
 addiction, 136
 cyber-bullying, 136
 distraction, 136
 privacy, 136
 sleep disruption, 136
 spreading of misinformation, 136
 texting, 136

Sociedad Quimica y Minera de Chile (SQM), 189
Socioeconomic differences, 30
Software-defined vehicles (SDVs), 64–66, 151–153
Solar capture, 250
Solar cells, 259
Solar power
 right-sizing
  battery, 266–270
  vehicle, 262–265
 solar panels, 258, 260
 solar photovoltaic modules, 261
 Teijin Limited, 257
 Toyota Prius PHV, 256, 257
Solar-powered charging stations, 95
Solar-powered vehicle prototypes, 288
Solar radiation, 259
S&P Global Mobility, 172
Sport utility vehicles (SUVs), 11
Star Trek, 135
Steering controls, 45
Structural reinforcement, 13
Sub-Saharan Africa
 motorcycle taxi drivers, 55
 observations from, 32–37
Substantial energy, 92
Supply chain, 143
 risks, 143
 security, 143
Sustainable approaches, 297
Sustainable strategy, 198–200
Systems thinking, 139–140

T

Taxi and For-Hire Improvement Fund, 243
Technological disruption, 139
Technology leadership, 2
Teetor, Ralph, 18
Teijin Limited, 257
Teleoperator, 114
Tesla Model 3, 54
Tesla Model S, 52, 54, 75
Tesla Roadster, 49

Tesla Supercharger, 60, 93
Thermal management, 90
Thermoelectric materials, 251
Thermoplastics, 300
Total addressable market (TAM), 324
Toyota, 121, 187, 253
Toyota Prius, 45, 47
Traction control system (TCS), 18
Traditional automakers, 4, 195–200
Traditional (greenfield) auto manufacturing plant, 306
Traditional automotive E/E architecture, 107
Traditional automotive factories, 305
Traditional OEMs, 199
Traffic fatalities, 157
Transformative technologies, 140
Transit-oriented development (TOD), 212, 214, 215
TransMilenio BRT system, 216
Transport for London (TfL), 314

U

Uber, 77, 78, 115
Ultra-fast charging, 90, 92, 93
United Arab Emirates (UAE), 194
United States (US)
 auto production share, 12
 average age of vehicles, 173
 commercial vehicle market sales, 236
 traffic fatalities, 157
 vehicle affordability, 172
United States Department of Transportation (USDOT) statistics, 157
Universal e-kit, 280
Urban congestion, 238
Urban design for mobility
 Barcelona's superblocks, 213
 bicycle parking, in Copenhagen, 214
 BRT, 216
 city policy, 211

equitable access to transport options, 217

free Wi-Fi, 219

mixed-use zoning, in urban areas, 214

neglect of active transportation, 212

parking space, 209

people-centric design, 213

Singapore's electronic road pricing, 219

strategically managing parking, 215

suboptimal urban mobility, 210

TOD, 214, 215

Tokyo's world-class rail system, 217, 218

urban planners and transport officials, 209

wide highways and large parking lots, 211

zoning laws and building codes, 214

Urbanization, 26–28

Urban mobility, 104, 299

Urban planning, 200, 201

V

Vancouver, 212

Vehicle affordability, 170–173

Vehicle bulging, 13

Vehicle construction, 303

Vehicle design and development

ACE platform "reference design," 273–278

car-free and/or low-speed zones, 270–273

e-kit ACE platform, 279–283

nature's lessons for future vehicle design, 252–256

reducing vehicle energy demand, 248–252

solar power, 256–270

Vehicle materials and manufacturing

circular economy, 297–301

economic benefits and supportive government policies, 314–323

localized production, 305–313

natural materials, 301–305

opportunity for automakers, 323–326

paradigm shift, 293–297

Vehicle miles traveled (VMT), 104

Vehicle ownership, 27

Vehicle performance, 16

Vehicle-related crashes, 154

Vehicle segment, 12, 13

Vehicle's suspension system, 254

Vehicle subscription service, 231–236

Vehicle-to-cyclist (V2C), 127

Vehicle-to-everything (V2X), 73, 124–128

Vehicle-to-grid (V2G), 95, 96

Vehicle-to-home (V2H), 95, 96

Vehicle-to-infrastructure (V2I), 125, 127

Vehicle-to-load (V2L), 95

Vehicle-to-pedestrian (V2P), 127

Vehicle-to-vehicle (V2V) communications, 73, 96–98

Vehicle usage, 7–9

automobiles, overengineering, 11–16

average annual mileage, 8

average traffic speed, 8

logical vehicle, 9

Maslow's hierarchy of needs, 9–11

national average annual mileage, 8

passenger-to-vehicle-weight ratio, 7, 8

prevalence of car culture, 8, 9

US average vehicle occupancy, 7

Vertical integration, 312

Vicious circle, 12

Vietnamese Government, 193

VinFast, 193

W

Wamae, Lincoln, 84, 85

Waymo's robotaxi fleet maintenance center in Phoenix, Arizona, 233

Wheelchair accessible vehicle (WAV) service, 163

Wheel motor, 101, 104, 105

Wi-Fi, 136

Wireless communications, 126, 221

Wireless EV charging (WEVC), 268, 269

Wood, 302

World Harmonized Light Vehicles Test Procedure (WLTP)-type approval conditions, 266

X

Xinjiang Uyghur Autonomous Region (XUAR), 141

Xpeng, 194

Y

YOYO XEV, 311

Z

Zambia-based Zambikes, 303

Zambikes, 85

Zero-emissions targets, 100

Zero-emissions vehicles (ZEVs), 99

Zipcar, 76

Zipline International, 241

About the Author

Dr. Christopher Borroni-Bird

Courtesy of Motor Trend

Dr. Christopher Borroni-Bird is an expert on sustainable and affordable mobility for ALL the world's people. This builds on his 25-year career leading advanced mobility initiatives at several major companies and universities as well as volunteering in sub-Saharan Africa.

With his consulting business, Afreecar, he is a senior advisor on future mobility to McKinsey and EVage (an Indian commercial EV startup) and is also sought out to provide independent technical due diligence on future mobility technologies and companies in vehicle autonomy, connectivity and electrification. Separately, Afreecar is also developing an e-kit that power assists non-motorized vehicles while also transforming them into mobile power sources, which has many applications in both the developing and developed worlds.

He served as Waymo's Chief Engineer for Future Vehicle Programs and has been a Research Scientist at MIT Media Lab. From 2012 to 2017, he was Qualcomm's VP of Strategic Development, responsible for wireless automotive solutions. Prior to this, he led GM's Electric Networked Vehicle (EN-V), the world's first drivable vehicles to demonstrate today's accepted vision of future mobility (and extensively deployed at the 2010 Shanghai World Expo). He led GM's Autonomy, Hy-wire, and Sequel "skateboard" electric vehicle concepts, now widely adopted by the Auto Industry, and has 50 patents. He is co-author of "Reinventing the Automobile: Personal Urban Mobility for the 21st Century," with Larry Burns and the late Bill Mitchell (MIT Press, 2010). Before joining GM in 2000, he led Chrysler's gasoline fuel cell vehicle development. He was inducted into the Automotive Hall of Fame as a Young Leader in 2000, was one of Automotive News' inaugural "Electrifying 100" in 2011, and was profiled by Motor Trend/Blackberry in the "Coding the Car" documentary and book series in 2023.

www.ingramcontent.com/pod-product-compliance
Lightning Source LLC
Chambersburg PA
CBHW040107100526
44584CB00029BA/3820